D1090408

OPEC
SUCCESS AND PROSPECTS

COUNCIL ON FOREIGN RELATIONS BOOKS

Founded in 1921, the Council on Foreign Relations, Inc. is a non-profit and non-partisan organization of individuals devoted to the promotion of a better and wider understanding of international affairs through the free interchange of ideas. The membership of the Council, which numbers about 1,600, is made up of men and women throughout the United States elected by the Board of Directors on the basis of an estimate of their special interest, experience and involvement in international affairs and their standing in their own communities. The Council does not take any position on questions of foreign policy, and no person is authorized to speak for the Council on such matters. The Council has no affiliation with and receives no funding from any part of the United States government.

The Council conducts a meetings program to provide its members an opportunity to talk with invited guests who have special experience, expertise or involvement in international affairs, and conducts a studies program of research directed to political, economic and strategic problems related to United States foreign policy. Since 1922 the Council has published the quarterly journal, *Foreign Affairs.* From time to time the Council also publishes books and monographs which in the judgment of the Committee on Studies of the Council's Board of Directors are responsible treatments of significant international topics worthy of presentation to the public. The individual authors of articles in *Foreign Affairs* and of Council books and monographs are solely responsible for all statements of fact and expressions of opinion contained in them.

OPEC

SUCCESS AND PROSPECTS

by

DANKWART A. RUSTOW

and

JOHN F. MUGNO

A Council on Foreign Relations Book
Published by
New York University Press · New York · 1976

Library of Congress Catalog Card Number: 75-29526
ISBN: 0-8147-7369-9

Library of Congress Cataloging in Publication Data

Rustow, Dankwart A
OPEC : success and prospects.

Includes bibliographical references.
1. Organization of Petroleum Exporting Countries.
I. Mugno, John, joint author. II. Title.
HD9560.1.066R87 382′.911 75-29526
ISBN 0-8147-7369-9

Manufactured in the United States of America

Preface

Not since the launching of the first Soviet sputnik in 1957 did Western powers feel as vulnerable as they did in the face of the Arab oil embargo of the autumn and winter of 1973-74. The sense of crisis was deepened and prolonged by the steep price increases decreed unilaterally by members of the Organization of Petroleum Exporting Countries (OPEC). But while there has been alarm over shortages of oil, there has been no shortage of analysis and commentary, whether by political leaders or by journalists, economic analysts, and academic specialists in international affairs. Much of this fast-growing literature has addressed itself rather single-mindedly to the question, What is to be done?

How can the United States make itself independent of energy imports? What can consumer countries do to break the power of the producer cartel? How can the world's financial system cope with the sudden displacement of accustomed flows of foreign exchange? Where can oil be found outside the sphere of OPEC control? How quickly can substitutes be developed for petroleum as a source of energy? What countermeasures can be devised against the possibility of another embargo? Such are the questions that have been posed with an understandable sense of urgency.

This essay addresses itself mainly to what seems to us a prior question. Instead of starting out to ask, What is to be done? it asks, What is happening to us? It focuses on the rise, slow at first, then meteoric, of the Organization of Petroleum Exporting Countries as a redoubtable force on the world scene. In recounting the successes of OPEC and assessing its prospects, the authors have inclined to the simple hypothesis (still unfamiliar to many American observers) that the leaders of the oil-producing countries, like other actors on the international political scene, act from motives of enlightened national self-interest, and that they have exploited rather skillfully a favorable conjunction of circumstances. In other words, we have assumed that no group of governments comes by $100 billion a year in a fit of absentmindedness.

As we pursued our investigation we concluded that there is indeed a logical pattern in OPEC's strategy and tactics—a plan that was set down as early as 1968 in OPEC's Declaratory Statement of Principles, and that has been implemented step by step as opportunity offered.

Our conclusion, in brief, is that OPEC has greater strength and hence better prospects than many observers, especially from the United States, have credited it with. It follows that many of the confident statements, heard so frequently in 1974, that OPEC like any other cartel would soon break up, or that the question was not whether the price of oil would come down but when, were based more on wishful thinking than on careful analysis of the international economic and political factors. Needless to say, such overconfidence on the part of OPEC's critics can only serve further to strengthen its own position.

The OPEC revolution has profound implications for the current and future international scene, and we have therefore tried to deal briefly with such themes as the recycling of petrodollars, the prospects of a "new international economic order" based perhaps on a whole new set of OPEC-like commodity cartels, the negotiations among consumer countries leading to the creation of an International Energy Agency, the diplomatic and military situation in and around the Middle East, and others. But our focus has always been on OPEC itself, on the question of cohesion among its member states, and on the external circumstances that determine its past and future success.

OPEC, it seems to us, has derived much strength and resiliency from sticking closely to its one dual task: that of helping assert the full sovereignty of its member governments throughout the Third World over their subsoil hydrocarbon resources, and of securing maximum monetary returns for them in the medium term from their petroleum exports. This has proved to be a dual aim upon which leaders of "revolutionary" juntas and "conservative" Shahs, sheikhs, and kings can readily agree as a basis for coordinated action.

Not long ago a writer for *Fortune* undertook a series of visits to OPEC's headquarters at Vienna and to the capitals of some of its member states. On his return, he noted with some surprise that "Practically the only action they ever fully agree on is raising the price of crude oil." This, of course, means that OPEC is nothing more and nothing less than a successful cartel.

Our narrative and our analysis are supported by detailed references to other literature and by a sizable statistical appendix, and it is our hope that these materials will be of value even to those readers (and there will be many) who may question our conclusions. International petroleum politics is perhaps the single most controversial theme on the world scene, and it has seemed to us appropriate that we should not ask our readers to take on faith the facts and figures on which our conclusions rest, but rather inform them in detail of the sources on which we have relied.

A brief preliminary draft of this study was presented in February 1975 to the first session of the "Survey Discussion Group on OPEC," organized by the Council on Foreign Relations. Walter J. Levy and John C. Campbell served as cochairmen of the group and Helen Caruso as rapporteur. Successive drafts have greatly benefited from critical comments by Messrs. Campbell and Levy and Ms. Caruso as well as by David B. Elkin, Peter Kenen (Princeton University), John H. Lichtblau (Petroleum Industries Research Foundation), George T. Piercy and Gerald Pollack (both of Exxon Corporation), and H. David Willey (Federal Reserve Bank of New York). Mr. Lichtblau and his associate at Petroleum Industries Research Foundation, Lawrence J. Goldstein, gave invaluable help in directing us to sources of statistics, and Messrs Piercy and Pollack and their associates at Exxon clarified important factual and statistical details.

It is a pleasure to acknowledge the financial support that John F. Mugno received for his portion of our research from the Ralph Bunche Institute on the United Nations at the Graduate Center of the City University of New York, directed by former Ambassador Seymour Maxwell Finger. Randi H. Crawford, a Ph.D. candidate in political science at CUNY, gave invaluable help in checking factual and statistical details and in editing the final manuscript. Karen Asakawa undertook the preparation of successive drafts of the typescript with rare competence and cheerfulness, continuing a close and valued association of more than eight years with the senior coauthor. In the final preparation of the book for publication it has been a privilege to be working with persons of such competence and good sense as Grace Darling Griffin and

Robert Valkenier of the Council on Foreign Relations and Despina Papazoglou of New York University Press.

The authors alone assume full responsibility for the facts and judgments presented throughout this book.

January 1976

Dankwart A. Rustow
John F. Mugno

Contents

[ONE]

The OPEC Revolution

The thirteen countries banded together in OPEC (Organization of Petroleum Exporting Countries) have posed a challenge to the international system that is without close precedent. Previous threats to the established balance have been military, or, as with the French and Russian Revolutions, ideological; OPEC alone is putting to the test the time-honored notion that economic forces are the determinants of history. Earlier challenges were continental, carried from Paris to the Pyramids and Moscow, or from Berlin to Stalingrad and el-Alamein. OPEC's challenge, originating in capitals such as Caracas, Tripoli, Riyadh, and Tehran, strained relations between the United States and Europe; endangered the world's banking and currency systems; prompted scores of governments to resort to novel measures of economic control; conferred new prestige on rulers such as Mohammed Reza of

Iran and Faisal of Saudi Arabia; reaped windfall profits for the Soviet Union and Canada; slowed economic growth in the industrial countries; imperiled development in many countries of Asia, Africa, and Latin America; and emboldened others to contrive cartels for primary commodities, ranging from bauxite to bananas. Rarely has an international crisis been so truly global or so pervasive in its effects.

Yet OPEC was founded less than sixteen years ago, played a rather insignificant role for a decade or so, and did not launch its revolution until the early 1970s. The interests it challenges are powerful, and its own cohesion will have to undergo further tests. No one can yet tell, therefore, how durable the international petroleum revolution will prove. The present essay will trace the origins of OPEC and the evolution of its program, assess the political and economic changes in the world petroleum market that provided the backdrop for OPEC's challenge, review the steps by which OPEC governments wrested control from the multinational companies and forced a fivefold increase in price, survey the responses of the oil-importing countries, and, on the basis of these, assess the current and future prospects of the organization.

Origins and Aims

When representatives of Iran, Iraq, Kuwait, Saudi Arabia, and Venezuela assembled in Baghdad on September 14, 1960, to found the Organization of Petroleum Exporting Countries, their immediate concern was with the pricing practices of the major international oil companies. Although aggregate revenues of the five governments from the export of petroleum had risen from $570 million in 1950 to $2.2 billion in 1960, income per barrel had risen at first, but declined steadily since

1957 (Iraq, Saudi Arabia) or 1958 (Kuwait, Iran, Venezuela). Saudi Arabia (1958, 1959), Venezuela (1959, 1960), and Kuwait (1959) even had experienced declining revenues not only per barrel, but also absolutely.[1] The producing countries thus were keenly conscious that their oil revenues, accounting for most of their government budgets, depended on prices set by a tightly interlocking group of companies with headquarters in the United States or in Europe.

It was the companies that decided where and when to prospect for oil; it was the companies again that determined how much oil to produce once it was found; and it was the companies that, in the light of their assessment of the market, set the price at which it would be sold. Although the companies operating in each country were legally distinct, there was a pattern of interlocking ownership: British Petroleum (BP), Shell, Mobil, and Exxon were part-owners of the three companies operating in Iraq; Exxon, Texaco, California Standard, and Mobil jointly owned Aramco, which operated in Saudi Arabia; Exxon, Shell, and Gulf, through affiliates, operated separately in Venezuela; Gulf and BP jointly held the concession in Kuwait. Later, all seven of them held shares in the Iranian consortium that replaced BP after the crisis of 1951-54. In the world at large in 1949/50, these seven companies controlled 65% of proved reserves of petroleum outside the Soviet bloc, 55% of its production, 57% of all refinery capacity and major pipelines, and, through ownership or long-term leases, at least 67% of all privately owned tanker space.[2]

This structure of horizontal and vertical integration implied that the prices charged for crude petroleum loaded at ports on the Persian Gulf or the Caribbean Sea represented little more than bookkeeping transactions between one company and its parent, affiliate, or sister company—transactions that through appropriate discounts could be readily adjusted so as to increase the ultimate income of the companies and to reduce the tax and royalty income of the country.[3] In 1959 and 1960, moreover, the companies began to cut prices all along the line,[4] and there was no telling what further cuts might be prompted by their efforts to compete with coal in the rapidly growing energy markets in Europe and Japan and to ward off the competitive inroads of "independent" companies outside the group of the seven.

Quarrels between producing governments and their concessionaires over the amount of payments to governments did not spring forth full blown after World War II. In Iran, for example, the 60-year D'Arcy concession of 1901 was superseded when the government, having become aware of the importance of oil to the British navy during the First World War, expressed dissatisfaction with its revenues, and insisted on renegotiation. Iran concluded a new agreement with the company (now named Anglo-Persian, and later Anglo-Iranian Oil Company) in 1920, but the government soon was perturbed that, during several years in the 1920s, production had increased but Iran's revenue had gone down.

In 1931 the government unilaterally cancelled the concession, citing the failure of the company to open its financial records to inspection. Moreover, as the Iranian Foreign Minister complained, the officials who negotiated the 1920 concession were not primarily concerned with the country's welfare–one of them, Armitage-Smith, had been sent as a financial adviser to the Persian government by the British government, the holder of a majority of Anglo-Iranian's shares. A new agreement was concluded in 1933, guaranteeing the government a minumum royalty per ton; until World War II, the two parties disputed whether the royalty should be calculated according to the English "long" ton (2,240 pounds) or the 2,000-pound "short" ton.[5]

There had been earlier attempts to coordinate policies among the governments of oil-producing countries. Venezuela had put out feelers to Middle East governments as early as 1947. Iraq and Saudi Arabia in 1953 had concluded an agreement calling for exchange of information and periodic consultation regarding petroleum. And an Arab Petroleum Conference in Cairo in 1959 adopted a resolution insisting that any changes in posted price should be discussed with the government of the producing country.[6] The major initiatives in this Latin American-Middle Eastern rapprochement that was to form the basis of OPEC were taken by the Venezuelan Minister of Mines and Hydrocarbons, Juan Pérez Alfonzo, and the Saudi Petroleum Minister, Sheikh Abdullah Tariki.

Yet the overwhelming reality of the 1950s had remained cooperation among the companies and division among the countries, a contrast revealed nowhere as starkly as in the Iranian crisis of 1951-54. Mossadegh, by a stroke of the pen, had proclaimed the nationalization of the Iranian assets of British Petroleum (the former Anglo-Iranian). But in view of the companies' control of tankers and refineries, the government found itself unable to market any substantial quantities of Iranian oil. Furthermore, although Iran in 1950 had accounted for nearly one-fifth of the world's petroleum exports, the shutdown there did not result in any shortage. The companies increased production rapidly in Saudi Arabia, Iraq, and, notably, Kuwait (where BP had a half-share of the concession), so that production in the Middle East as a whole soon reached new peaks. And exploration kept well ahead of production: 25 times as much oil was added to the companies' inventory of "proven reserves" in the Middle East in 1951-54 as had been produced in Iran in the entire period from 1913 to 1950.[7]

Although the British government decided against military action, the cruiser *Mauritius* made a demonstrative appearance in the Persian Gulf, and the London authorities for a time suspended essential exports and financial transfers to Iran.[8] Mossadegh's dramatic action turned out to be ill-prepared and premature. Even before his government was overthrown with an assist from the CIA, the oil companies had thwarted his economic attack by marshalling their alternative resources across the Persian Gulf.

Nor did the founding of OPEC lead to any immediate reversal of relations between companies and countries. Conferences of representatives of the governments were held at the ministerial level twice a year. (The first OPEC meeting of heads of state was held only in 1975.) A small secretariat was operating at headquarters in Geneva and then in Vienna. Conferences were held there or in the member countries' capitals and required unanimity for their resolutions—but presumably a great deal of informal business was handled as well. New members were admitted from among Third World countries with sizable petroleum exports until in 1975 the membership had grown to 13.

Some of OPEC's founding governments favored the classic cartel strategy of curtailing production so as to drive up prices, and hence government royalties and income taxes.[9] But this would have required allocating production quotas among countries—a task that would have been sure to shatter the fragile unity of the infant organization. It would also have run directly counter to the policy of the companies, which had always claimed exclusive authority over decisions on volume of production and which were just then engaged in a major campaign to expand their markets.

It was not until mid-1962 that the fourth OPEC conference devised a tactic which, after three more years, was to lead to the first modest success. It consisted in bargaining with the companies, according to a concerted plan, over the accounting details on which the governments' revenue per barrel depended. Of the three substantive resolutions adopted at the time, one recommended that prices be brought up to their pre-August 1960 levels; and another that "royalty payments . . . shall not be treated as a credit against tax liability." [10] And even this tactic was only partially successful. Posted prices remained steady for an entire decade ($1.80 for Arabian light, which served as what later became known as "marker crude," i.e., the crude petroleum that provided the benchmark price for oil of other qualities or produced at other locations). The most that OPEC could claim to have achieved was to have prevented further cuts beyond the decline in real prices implicit in a steady nominal price.[11] On the other hand, the companies by 1965 agreed with each of the several countries not to claim the royalty as a credit against the 50% income tax.[12] Under these new formulas, the per barrel revenues in, e.g., Saudi Arabia went up from 78.7 cents in 1963 to 83.4 cents in 1966.[13]

This first partial success encouraged OPEC, at its sixteenth meeting in June 1968, to adopt a "Declaratory Statement of Petroleum Policy in Member Countries." The statement received little attention in the Western press and "was not taken seriously by the oil companies at the time," [14] but it constitutes in fact the single most important document embodying OPEC's long-range strategy as it has unfolded in the 1970s. It was foreshadowed by a speech in Beirut three weeks earlier by Saudi Arabia's oil minister Sheikh Ahmad Zaki al-Yamani, whose role in implementing the grand design has since become fully apparent.[15]

[8]

The ten-point Declaratory Statement included the following demands: (1) "Member Governments shall endeavour, as far as feasible, to explore for and develop their hydrocarbon resources directly." To the extent that they are unable to do so, contracts with foreign companies may be concluded, provided that the government retains "the greatest measure possible of participation in and control over all aspects of operations" and that "changing circumstances should call for revision of existing concession agreements." (2) ". . . the Government may acquire a reasonable participation on the grounds of the principle of changing circumstances." (3) There should be "progressive and more accelerated relinquishment of [any] acreage" of existing company concessions where exploration and production have not in fact occurred. (4) Posted prices "shall be determined by the Government" and adjusted against declining monetary values. (5) "The Government may, at its discretion, give a guarantee of fiscal stability to operators for a reasonable period of time." (6) Such guarantees are to be renegotiated if for any year just ended the company is found to have realized "excessively high net earnings after taxes." The remaining points claim for governments the right (7) to set reasonable standards of accounts to be kept and information to be furnished by the companies, (8) to formulate "the conservation rules to be followed," (9) to exercise full jurisdiction in its "competent national courts" in any disputes with the companies, and (10) to invoke against the companies the rule of "the best of current practices" for such matters as incorporation, labor relations, royalties, taxes, and property rights.[16]

OPEC had formulated a grand charter that, if implemented, would transfer from companies to countries full control at the upstream end of operations, leaving the companies in a merely technical, though adequately remunerated, capacity. On the assumption that OPEC governments could concert their setting of posted prices as smoothly as had the companies, it would mean transforming the market arrangements of the multinational companies into a tight cartel of a handful of countries. And since the desire for higher revenues was the very stimulus that had called OPEC to life, it was plain that the prices imposed by OPEC would be substantially above those charged by the companies. How soon and how far OPEC would be able to implement its program depended on political and economic circumstances and OPEC's skill in exploiting them.

Companies and Governments

In the quarter-century after World War II, the economies of the non-Communist countries of the world, particularly the United States, Japan, and Western Europe, experienced an upsurge of unprecedented magnitude. International trade expanded fivefold, the world's gross domestic product tripled, as did energy consumption.[17]

The proportion of this burgeoning energy market that petroleum would command depended very largely on its price, and the price in turn depended on an intricate combination of economic and political factors. The economic factors in themselves are straightforward. There is a lower limit to the potential price of oil, represented by the physical cost of production in the most favorable fields, plus costs of capital, transportation to markets, refining, and distribution: no one will long produce oil at a loss. And there is an upper limit representing the cost of producing alternative fuels (say, coal for furnaces and railroad engines, nuclear power for electricity, and synthetic petroleum from coal for automobiles): no one will long use a more expensive fuel if he can find a cheaper one. It is within this economically determined range of potential prices that political factors come into play—and in the oil business the potential price range is enormous and the political factors are potent. In practice, the political economy of oil usually reduces itself to a major contest of wits and of wills between oil companies, the governments of producing countries, and the governments of consuming countries.

In a world of global free trade and perfect competition, the price of oil would be near its lower limit and consumption at a maximum—as long, that is, as present sources of oil lasted or new ones could be found. In such a world, most of the petroleum consumed in the last quarter-century would have come from the Persian Gulf, where the cost of producing a new barrel of oil in the 1950s was between 12 and 22 cents, as against 39 cents in Venezuela and $1.51 in the United States.[18] Nor would there have been any danger of early exhaustion of the available petroleum: in 1955 the ratio of proven reserves to annual production around the Persian Gulf was as much as 106:1, as against a mere 12:1 in the United States.[19]

In a world of perfect monopoly, the price of oil would be at the upper limit drawn by available alternatives—or as close to it as the monopoly could raise it without provoking a reduction in consumption so severe as to reduce its total profit. Such a world can readily be imagined if we superimpose the political and corporate map of 1911 on the economic conditions of 1950. Imagine, that is, an undivided Standard Oil Company, already in control of most petroleum production in the Western Hemisphere, obtaining a concession from the Ottoman Sultan covering oil production in Mosul, Kirkuk, Kuwait, al-Hasa, and Libya, and one from the Shah covering Khuzistan. In such a situation, too, most of the oil consumed in the world would come from the Persian Gulf, although total consumption, at prices imposed by a near-perfect monopoly, would be less.

The lower of these two limits is not difficult to estimate: it would be around 12 cents a barrel, plus a few pennies of profit, f.o.b. Persian Gulf, or (adding costs of transportation) around $1.30 in the United States, and slightly less in Europe.[20] The upper limit is impossible to estimate with precision, for substitutes to run the world's automobiles and planes on fuel other than petroleum are not currently available at any price, and—happily for the world's consumers—the elasticity of the demand for oil at prices above $11 a barrel f.o.b. Persian Gulf has not yet been tested. But it is not unreasonable to guess that in the short run this upper limit lies well above $20.[21]

The stakes of the political struggle for oil have been how to allocate this enormous differential between 12 cents and $20-plus a barrel—a prize that with increasing consumption in the non-Communist world since 1960 has grown from about $50 billion to about $200 billion per year. The struggle was being fought on a stage already crowded with other contestants, notably major oil companies, independent oil companies, and consumer country governments, when OPEC in the 1960s made its undramatic entry.

The relevant policies of the consumer country governments in the United States and Western Europe may be described briefly as imperialism abroad and protectionism at home, though each of these attitudes took a number of complex and subtle forms. In 1919-20, Great Britain and France established their political hegemony in the Middle East and, as a natural byproduct, agreed to share, three to one, the petroleum concessions in the former Ottoman territories. United States pressure for an "open door" by 1927 changed this ratio into one of 2:1:1 in favor of Britain, the United States, and

France.[22] In Kuwait, British and United States interests shared the concession. In Saudi Arabia, only American companies were concerned. And in Iran, the former British concession in 1954 was converted into a consortium where British and United States interests each held 40%, and Royal Dutch-Shell and Compagnie Française de Pétroles (CFP) the remaining 20%. In 1967, the overall shares of Middle East production were American 57%, British 28%, British-Dutch 7%, and French 6%.[23]

The French in 1946 withdrew from their mandate in the Levant, and the British phased their withdrawal over the next quarter-century: Suez 1954-56, Jordan 1956, Iraq 1958, Kuwait 1961, Aden 1968, and the Emirates 1971. American power was in the ascendant, but was exercised preferably through subsidies rather than through colonial occupation. The pre-1970 situation of Middle East oil was a company regime supported by British military and American economic power. The American preference for financial subsidies over other forms of involvement was evident in aid programs to Israel (including large private donations), Turkey, Iran, and more sporadically to Egypt, Saudi Arabia, Jordan, and other countries. It was equally apparent in the policy of the Internal Revenue Service that allowed American companies to claim income taxes paid to foreign governments as full credits against American federal taxes on their foreign operations.

The foreign tax credits are of fundamental significance because they made possible the conclusion of the so-called fifty-fifty agreements, and these in turn furnished the framework for the financial relations between the multinational oil companies and their host countries in the 1950s and 1960s. Fifty-fifty agreements were concluded in Venezuela (1948), Saudi Arabia (1950), Kuwait (1951), Iraq (1952), Iran (1954), and elsewhere. As the phrase implied, they provided for an equal division of profits between companies and governments, and since the prevailing royalty rate was one-sixth in Venezuela and one-eighth elsewhere, this meant a tripling or quadrupling of government receipts. To make the new arrangement acceptable to the companies, however, the additional payments took the form of income tax rather than royalty, so that they became, for United States tax purposes, not a mere business expense but rather a full credit against the company's U.S. income tax on its foreign operation. The tax provisions in the relevant European countries (Great Britain, the Netherlands, and France) were different in detail but by the mid 1960s came to be analogous in their results. "Effectively, therefore, the shift to 50:50 . . . represented largely a transfer of tax revenue from the American and the British treasuries to those of the Middle East [and Venezuela]. The companies became the channel for this transfer . . . "[24] In the judgment of one observer, this new tax structure created "a strong identity of interests between producing governments and the companies," and for United States policy makers concerned with the Middle East provided "a seductive alternative for a more politically controversial direct foreign assistance program" to Arab countries and others.[25] Specifically, Aramco's payments to Saudi Arabia in the first full year of the fifty-fifty agreement rose by over $50 million, whereas its United States income tax gradually dwindled to the vanishing point.[26] Note that from the mid-1960s onward, when royalty payments were added to, rather than deducted from, income tax payments to OPEC countries, the companies began to accumulate vast excess tax credits. These details of tax-accounting procedure meant that the division of production profits, which in 1948 had been approximately 63:37 in the companies' favor passed the 50:50 mark in 1955-56 and became approximately 70:30 in the governments' favor by 1970.[27]

Consumer government protectionism remained just as

important a factor in the triangular relations between importing countries, companies, and exporting countries. The American major companies had set a precedent by using Galveston, Texas, as the "basing point" for their international prices—making sure that oil from the Persian Gulf would get as high a price in the American market as their domestic production, and a correspondingly higher one elsewhere in the world—thus enabling the companies to recoup their Middle Eastern investments every two years or so.[28] Somewhat later, the United States government protected high-priced domestic oil more directly through the quota system first introduced in 1959 and maintained, with variations, until the spring of 1973. This policy at the time of its inception was justified by spurious arguments from national security, but in fact it made Americans forego oil imports when they were cheap and without strings, and depend on them when they became dear and politically tied. S. David Freeman has sarcastically but aptly called it a "Drain America First" policy.[29] European governments achieved a similar result through excise taxes on petroleum products that often exceeded the cost of importing and processing—the aim in this case being to protect the high-priced coal industry and not antagonize the powerful coal miners' unions.[30]

The governments of oil-producing states played the second major role in the oil drama. Of these, Venezuela had long been the most active: it had been in the oil business longer than most and had never been as heavily under foreign influence as the Middle East. As noted earlier, Venezuela pioneered the "fifty-fifty" agreements, which added more to the treasuries of oil-producing governments than any measures implemented in OPEC's first decade; and it repeatedly took the initiative in associating with its fellow oil exporters in the Middle East. By the 1960s, Middle Eastern and North African governments also made their contributions. Thus Libya transferred to the Middle East the earlier Venezuelan pattern of awarding not a single country-wide concession, but different ones for separate bits of territory. Iran followed the same practice in awarding offshore drilling rights in its part of the Gulf, with stringent limitations on the number of years from award to exploration and from discovery to production, in default of which any concession would revert to Iran. By the 1960s a score of oil companies, singly or in combinations, were producing in Libya.

Earlier, a single, closely interlocking set of companies had dealt with a divided group of countries. OPEC tried to redress this disparity by overcoming the division among the countries. Libya's policy reversed the former situation fully by dividing the companies and thus allowing a single government to rule them. (Note also that for many of the newcomers among the oil companies, Libya was their only source of crude. Hence, unlike the majors, they were not inclined, in any dispute with the government, to face the threat of a shutdown.) Whether the initiative came from Venezuela, Libya, Iran, or the Arab countries around the Persian Gulf, or from OPEC itself, the constant aim was to expand the governments' control over production and to maximize the revenues accruing to them. These were the very principles embodied in OPEC's bold Declaratory Statement of 1968.

The oil companies that moved on this shifting political scene included not only the seven majors, which controlled the international oil industry until about 1950, but a score of other companies eager to cut into the majors' profits by offering better terms to Middle Eastern or North African governments. These new arrivals were of several kinds. Some were American integrated companies, such as Continental and Marathon, without previous foreign sources of production.

Some, such as Occidental and Ashland, were American "independents"—that is, refining companies with no previous production of crude petroleum at all. And some were companies, mostly state-owned or state-supported, from Italy, Japan, Spain, West Germany, and other countries. Most active in this last group was Italy's ENI, whose chief Enrico Mattei did not tire of battling the major oil companies. (Derisively, he called them the "Seven Sisters.")[31] Luckily for the seven majors, few promising concessions were left for the latecomers, except in the Kuwaiti-Saudi Neutral Zone and in Libya (and in Libya, the majors managed to obtain a sizable share of the production, after all).[32]

Other newcomers were the national petroleum companies of Middle Eastern countries, such as the National Iranian Oil Company, which undertook exploration, alone or in partnership with foreign companies, in areas not previously conceded or where the concession had lapsed. Still, despite this multifarious activity, the total share of the world's petroleum exports in 1970 produced outside of the seven major companies was no more than 18%.[33]

The major companies fended off the double onslaught of nationalism in producing countries and of competition from independents and others remarkably well for twenty years from 1950 to 1970. First, in coping with Mossadegh, they relied on the traditional hegemony of the Western powers in the Middle East and on their world-wide alternative resources, which allowed them to shut down operations in any country that proved unreasonable. Later, they came to rely on tax regulations at home, which enabled them to multiply payments to governments abroad at no initial cost to themselves. And finally they met mounting pressures from OPEC and newcomers by allowing the price of oil to decline in relation to that of competing fuels, thereby vastly expanding their markets.[34]

Because American oil import quotas provided an effective protectionist barrier, the major expansion took place in Europe and Japan where there was a steady rise in the cost of labor-intensive coal from ever more meager seams. All in all, oil in international trade increased more than sixfold between 1950 and 1970; payments to producing governments leapt more than twelvefold from $570 million to $7 billion; and company net income from foreign production still advanced steadily, quadrupling from about $800 million to $3.3 billion in those two decades.[35] In retrospect, foreign tax credits and market expansion emerge as the most important elements in the companies' grand strategy.

But the political and economic circumstances of the international oil industry were changing rapidly, just as the major oil companies were reaching the end of their two avenues of retreat. The British government, as part of its new low profile East of Suez, evacuated its last naval and military units from the Persian Gulf in 1971. Appeals and financial offers from some of the local rulers ("Who asked them to leave?" a perturbed sheikh of Dubai inquired of the correspondent of the London *Times)* proved unavailing, and so did suggestions that the United States (still deeply enmeshed in its unfortunate venture in Vietnam) fill the gap. In the dispute between BP and Mossadegh in the early 1950s, the British government had kept units of its navy and air force at hand. Now the prospect of Western gunboats or marines would be far more unlikely. Further market expansion would be difficult, since Japan's shift from coal to oil was virtually complete, and Europe's largely so, whereas the United States clung to its quotas until the spring of 1973. The "fifty-fifty" arrangements in Venezuela and the Middle East had virtually wiped out the companies' back-home tax liability on their foreign operations. And companies, of course, can write off no more than 100% of their taxes, just as customers can switch from coal to oil for no more than 100% of their consumption.

A third avenue of escape—to look for oil outside the politically volatile Middle East—did not lead very far; for some of the best new finds—Libya, Algeria, Abu Dhabi—were in other Arab countries, and all of them, until the late 1960s, were in the Third World—Nigeria, Indonesia, Ecuador, Gabon. Within a few years of going into commercial production, each of these had joined the ranks of OPEC. The organization's share of world oil exports remained remarkably steady at 82% to 84% throughout the 1960s.[36]

By 1970 the change was complete. In view of the West's political and military withdrawal from the Middle East, company resistance against government demands would have been futile. (Professor M. A. Adelman, the leading academic petroleum economist in the United States, has severely criticized representatives of the Department of State for advising oil multinationals to accede to producing-country demands. But this was not so much an independent development, as Adelman assumes, as an acknowledgment of the new situation.[37]) And since the companies had run out of additional tax credits to claim and out of new markets to conquer, they could not meet any new financial demands by OPEC without cutting into their own profits or raising prices to their customers. Naturally they chose the second option.

In short, the companies, in any new confrontation with OPEC or its member countries, could be expected to yield, and to pass along any financial pressure to their customers. They were not exactly relegated, as the chairman of BP put it, to being a vast "tax-collecting agency" for OPEC,[38] since they continued in charge of the world-wide transport, refining, and marketing of oil and made profits somewhat beyond a tax collector's normal salary. But they had lost their control-

ling position at the upstream end from which their other powers once had flowed. And in view of Japan's and Europe's mounting (and America's prospective) dependence on oil imports, any concerted drive by OPEC for higher revenue would be sure to have repercussions throughout the entire industrial non-Communist world.

Whether the strategists of OPEC and its member countries fully appreciated all the factors just listed is hard to tell. But several things were certain: that objective circumstances had fundamentally changed in their favor; that OPEC countries were ready to exploit fortuitous circumstances (such as the temporary closing in 1970 of the pipeline from Saudi Arabia to the Mediterranean and the resulting temporary tanker shortage) so as to probe defenses on the other side; that they were ready to push ahead singly or in groups, using to best advantage such idiosyncratic factors as the penchant of Libya's President Qaddafi for fire-breathing harangues, the Shah's well-nurtured reputation of reasonableness, or King Faisal's image as a staunch anti-Communist supporter of the West; and that each new foray served to advance OPEC's grand design.

Cash and Control

In the pursuit of the grand goals of the Declaratory Statement of Petroleum Policy during the years since 1968, the chief function of the OPEC secretariat in Geneva and later in Vienna has been the coordination of information. At its periodic meetings OPEC has frequently passed resolutions expressing support for a member's individual efforts. Occasionally there have been unanimous statements on negotiating priorities, and from time to time OPEC itself has appointed a negotiating team—or perhaps, rather, given its blessing to teams that volunteered. Aside from these activities of the organization as a whole, different members have taken the lead at different times, and others have followed suit.

The intense round of negotiations of the winter of 1970-71 constitute a crucial turning point in government-company relations. Prior to this time—except for Mossadegh's abortive attempt at nationalization—the governments had pressed for larger payments, and on occasion had challenged the companies' accounting practices. But any changes had come about through negotiations, and the governments had not disputed the companies' control of the production process. If negotiations came to any impasse, the companies could rely on the implicit threat of closing down operations in a given country. In short, OPEC still was unsure of its own potential power. Third World countries were thought to be "less able to operate a successful cartel" than were Western businessmen.[39]

Libya's Qaddafi in September 1970 exploited the conjunction of increases in demand and expected decreases in supply to begin showing both OPEC governments and companies the errors in these assumptions.[40] He ordered production cuts which aggravated the developing shortage, and he used mandated production cuts and the threat of shutdown effectively against individual companies. Putting pressure first on the weaker independents, he raised posted prices by $.30 per barrel and increased the tax rate from 50% to 55%. The Persian Gulf countries, such as Iran and Kuwait, not only refrained from undercutting the Libyan position but also joined in the demands for higher prices and 55% tax.[41]

From this dramatic opening, a fairly routine pattern developed in the next five years in negotiations between countries and companies. From late 1970 to mid-1974, revenue increases and participation gains alternated as

OPEC's proximate goals; since mid-1974, the two strands have been woven together more closely. Typically, at the beginning of each phase, an OPEC conference set forth the goal to be achieved; subsequently one or another member took the lead; and finally all others would insist on gaining "the best of current practices."

Peregrine Fellowes, writing as early as 1971, described the strategy of the producing countries succinctly: "After cash must come control."[42] The tactics have been even more resourceful: more cash, more control, then more cash and yet more control, and then still more cash.

OPEC's first line of attack, following Qaddafi's lead, concentrated on augmenting government revenues. Its Caracas meeting of December 1970 explicitly recalled the 1968 Declaratory Statement and adopted as a minimum tax rate Qaddafi's 55%.[43] Venezuela at once raised its rate to as much as 60%, made that increase retroactive to the beginning of 1970, and henceforth claimed the right to set tax-reference prices unilaterally. The most dramatic negotiations took place two months later in Tehran, where 23 companies were confronting the governments of the Persian Gulf area, and where over two-thirds of OPEC's production was at stake. The companies, wary of Qaddafi's earlier tactic of picking them off one by one, had insisted on negotiating as a single group; but so had the governments. When conversations deadlocked, OPEC's conference reconvened and threatened to embargo all supplies of crude and refined oil of any company that would fail to accept the principle of the Caracas resolution—that is, a 55% tax rate.[44] The major companies consulted informally with the U.S. Department of State, which urged the companies to resolve the matter by negotiation. In the Tehran agreement of February 14, 1971, the companies accepted the 55% rate, an immediate increase in posted prices, and successive further increases spread over the next four years. As a result, the average government revenue per barrel of crude rose from $.87 in 1970 to $1.25 in 1971. Even without any increases in amounts of production, the Gulf countries would raise their annual income from about $4 billion to $6 billion at once, and to about $7.5 billion by 1975.

The Persian Gulf countries thus had obtained financial terms comparable to those extracted by Libya the previous autumn. But Qaddafi was not content to remain first among equals. Relying on the low sulfur content of Libya's crude and its proximity to major European markets, he insisted on a premium of nearly $1 above the posted prices set at Tehran, and similar markups were obtained by Saudi Arabia and Iraq (for the throughput of their Medterranean pipelines), Algeria, and Nigeria. (Nigeria had dramatically joined OPEC while the Tehran negotiations were in progress.)

The next target of OPEC countries was a government share in the ownership of producing companies. Here Algeria, embroiled in a dispute with France, acted first, in February 1971, by nationalizing all pipelines and gas fields and 51% of all French oil concessions. French efforts to organize a boycott proved unavailing, and before the end of the year, the dispute was settled mostly on Algeria's terms. Saudi Arabia and the neighboring Emirates of Kuwait, Qatar, and Abu Dhabi disavowed any intent of "nationalization" and instead pursued the plan of "participation"—that is, a gradual transfer of ownership from company to government. OPEC assisted by appointing a ministerial committee of these very countries to study and report on the matter. In August, Venezuela, ever proud to have been OPEC's chief initiator, enacted its own version of gradual takeover in a "Hydrocarbons Reversion Law" under which government ownership was to be complete in 1983. The ministerial committee opened negotiations on behalf of the four Arab countries in January 1972; OPEC came to its aid in March by threatening "appropriate sanctions"—presumably a cut-off of supplies—against any company that should "fail to comply with . . . any action taken by a Member Country in accordance with [OPEC] decisions . . ." [45] In October negotiations were concluded, with participation to begin at 25% in 1973 and to reach 51% in 1982.

In January 1973 the Shah of Iran announced that the 1954 consortium agreement would not be renewed when it was due to expire in 1979; in March (on Persian New Year's Day) he was able to announce the consortium's immediate nationalization in return for assurances of a continued 20-year supply of the companies' normal quotas of Iranian oil. The Kuwaiti legislature, basking in its newly won right to ratify agreements and noting that Iraq and Iran had gone the way of complete nationalization, rejected the 25% to 51% participation plan in September 1973 and instead insisted on an immediate 60%. By mid-1974 the other "participation" countries, including Saudi Arabia, matched this 60% bid. By year's end, Saudi Arabia went all the way, demanding a 100% takeover. Much of 1975 was spent in negotiating details of the settlement, Aramco's major concern being assurances of future supply for its four parent companies rather than compensation for past investments that had been recouped many times since.

Meanwhile, Libya had nationalized 50% to 100% of various foreign oil companies in its territory in a protracted process that began in December 1971 with the takeover of the BP portion of the BP-Bunker Hunt concession, in retaliation against alleged British connivance in Iranian occupation of three small islands at the mouth of the Persian Gulf and culminated with the 51% nationalization of the shares of eight European and American companies in September 1973. In

June 1972, Iraq nationalized the Iraq Petroleum Company's Kirkuk field, after the company's action in cutting production, and hence Iraq's revenue, had brought to a head the long-smoldering dispute over compensation for the North Rumaila concession (which Iraq had taken over in 1961 and later began to develop with the help of Soviet technicians). This time, too, OPEC proclaimed its solidarity, warning companies not to undercut the Iraqi position by increasing production elsewhere.[46] The Exxon, Mobil, and Gulbenkian shares of another IPC-affiliate concession, the Basrah Petroleum Company, were nationalized as a result of the Yom Kippur War. By the end of 1975, by means either of participation or of partial nationalization, all Middle Eastern and North African countries owned 50% to 100% of their petroleum production.[47]

The continuing depreciation of the U.S. dollar prompted OPEC in September 1971 to ask for a corresponding increase in posted prices. In an agreement signed at Geneva in January 1972 this increase was fixed at 8.49%, and it rose by another 5.7% in April 1973. But posted, or tax-reference, prices were rising on other occasions as well, in line either with the Tehran agreement or with the right claimed earlier by Venezuela to fix prices unilaterally. The government "tax take" on the barrel of oil, which at the beginning of 1973 had ranged from $1.47 (Kuwait) to $2.22 (Libya), by October 1 ranged from $1.72 (Kuwait) to $2.90 (Venezuela).

All earlier price increases were dwarfed by those of mid-October 1973, which raised the posted price of the marker crude (34° Arabian light ex Ras Tanura) from $3.01 to $5.12 per barrel, and by those of late December (effective January 1, 1974) which raised it to as much as $11.65—implying a "tax take" of $7.01.[48] The first round of increases accompanied the Arabs' use of their long-heralded "oil weapon" in the Yom

Kippur War. Oil shipments were embargoed to the United States, the Netherlands, and certain other countries; production was cut by as much as 25% at once and further cuts of 5% a month were threatened; and posted prices were raised by 70%. The embargo and production cuts were fully applied only by the Arab producers of the Gulf (Saudi Arabia, Kuwait, Qatar, Abu Dhabi), and somewhat more unevenly by Libya and Algeria. Iraq, while participating in the embargo, soon restored full production. (Among the major non-Arab countries, Venezuelan production remained steady, whereas that in Iran, Nigeria, and Indonesia increased slightly.) Still, the overall effect was a decrease of world oil supplies in international trade from 33 million barrels per day (mb/d) in September 1973 to 28.8 mb/d in November 1973.[49] Since the allocation of the shortage had to be handled by the oil companies, which to this day control the world's network of tankers and refineries, the effects were fairly evenly distributed. But an even reduction of oil imports meant a much more severe energy shortage in Europe and Japan (which depended on oil imports for three-fifths or three-fourths, respectively, of their total energy) than in the United States (where oil imports constituted only one-sixth of total energy consumption).[50]

All OPEC countries participated in the price rises—indeed, it was the Shah who pressed for the December round against the resistance of Saudi Arabia's oil minister Sheikh Yamani. And even non-OPEC members followed suit. Canada, which then still was exporting much of its oil from Alberta to the United States but importing roughly three-fourths of that amount from OPEC countries into Quebec and Ontario, fully matched the OPEC price increases. And the Soviet Union by early 1975 was reported to be doubling the price of its oil exports to other COMECON countries.[51]

The financial effects on consumer countries were world-wide, those hardest hit being the resource-poor countries of the Third World (such as India, Pakistan, Bangladesh, Sri Lanka, and most of the smaller countries of Africa) and those industrial countries (e.g., Italy and Great Britain) which were already in foreign payments difficulties before the petroleum crisis.

After the end of the embargo in the spring of 1974, OPEC governments concentrated once again on expanding their control over petroleum operations and on increasing their income per barrel. Increased control in 1973 and 1974, as we have seen, often took the form of government "participation," a feature that vastly complicated the applicable accounting procedures. For many of the countries throughout most of 1974, as many as four different prices were in effect: the "posted price," on which the companies' royalty and tax liabilities were calculated; the price at which the government sold some of its own royalty or participation crude to outsiders; the "buy back price" at which companies were committed to purchase the remainder of the government's share; and the price at which companies actually sold crude.[52]

This complex accounting process made possible a number of price increases without affecting the fictitious but widely publicized posted price. Indeed, OPEC at each of its quarterly meetings in 1974 decided to "freeze" the posted price. But a rise in royalty rates from 12.5% to 14.5% was decided in June, and by the end of 1974 most countries had raised the rate to 20%. A 5% rise in income tax was decided at OPEC's conference in September 1974, but by year's end, most countries had far exceeded that target—the tax rate standing at 59% in Ecuador, 61% in Nigeria, and 80% in the Arab Gulf countries. Saudi Arabia achieved another increase in income by raising its "buy back price" by 13 cents a barrel.[53] The net effect of all these arithmetic maneuvers was to raise the level of government revenue per barrel during 1974 by more than 40%.

Typically, the Arab countries around the Gulf moved in concert on both participation rates and tax and royalty rates, with Iran cutting through the maze of accounting intricacies by adjusting its tax level to the resulting tax-paid cost of crude. The other OPEC members have felt free to depart from that common pattern. For example, Indonesia broke the OPEC "freeze" by raising its official selling price to $11.70 in April and $12.60 in June; [54] and Ecuador led the increases in the tax rate—although it had to reduce its tax in mid-1975. Since Iran and the Arab countries on the Persian Gulf account for two-thirds of OPEC's output, their cohesion has guaranteed OPEC's price structure, and other exporters have been able to step out of line only so far as their closeness to markets (and sharply fluctuating tanker rates) permitted.

The last phase of the OPEC revolution, since the end of 1974, has been marked by moves to full government ownership and toward a simplified accounting structure. These moves were foreshadowed by the commissioning of two studies by OPEC's September 1974 conference, one for a new long-term price structure, and the other of ways and means of accomplishing full production control. In November, the Saudis hinted at selling their government oil to independent companies at a lower price than Aramco's weighted tax-paid cost—a move that led to a quick resumption of the long-stalled negotiations and, by December 6, induced Aramco to agree in principle to yield 100% control to the Saudis. These moves shortly were paralleled in Kuwait, Qatar, and Venezuela. OPEC's December conference also decided to simplify the accounting structure by abandoning the "posted price" and all the complex calculations based on it. Instead there was to be a single specified level of government return per barrel (e.g., $10.09 for the Saudi marker crude)—thus in effect adopting the system that Iran had followed for some time.[55]

When representatives of twenty-three Western oil companies signed the Tehran agreement with six Persian Gulf governments, they reluctantly conceded increases in the tax rate and in posted prices that, by 1975, would nearly double the companies' payments to governments. But they comforted themselves—or at least the audiences of their press conferences—by insisting that they had negotiated a stable arrangement under which their access to crude would be guaranteed and under which companies and governments could coexist in predictable fashion for five years. In fact the Tehran agreement ushered in a period not of stability, or even of orderly transition, but rather of breathtaking and unanticipated change. In the five years before Tehran, payments per barrel had risen only 16 cents; under the Tehran agreement they were supposed to rise by a total of 60 cents; in fact, they rose more than tenfold before the end of 1975. And whereas the companies before 1971 had determined levels of production and patterns of distribution, the OPEC countries henceforth claimed the right to limit production at their own discretion and (as in the embargo of 1973-74) to tell the companies where their crude petroleum could and could not be sent. If the OPEC revolution is to be pinpointed to a single date, the day is February 14, 1971.

By 1975, most of OPEC's ambitious agenda of 1968 had been carried out. No one any longer disputed the governments' full sovereignty over the subsoil hydrocarbon resources of their countries. The governments freely exercised their right to limit production for geological, economic, or political reasons. They could decide to whom the oil would be sold and how much they would receive for each barrel. And the governments or their national companies reserved the right to take over whatever portion of the process of production, refining, and marketing they could (or felt they should) handle.

Increases in government revenue per barrel, limitations on production, and political embargoes came to be unilaterally announced by the governments. On other matters—participation, nationalization, financial adjustments for declining currency values—there were negotiations with the companies. But a comparison with the Mossadegh crisis shows how thoroughly the bargaining situation had been reversed. In the early 1950s, British Petroleum had increased its production in Iraq, Kuwait, and elsewhere to make up for the shutdown in Iran; and the Iranian government, without financial reserves, was soon on the verge of bankruptcy. In the early 1970s, OPEC repeatedly made it clear that, in case of deadlock, a company would not be allowed to increase production elsewhere and might even find itself embargoed throughout OPEC. And this time, whereas the companies and the oil-consuming countries disposed of stored reserves of petroleum ranging from one to three months consumption, the major OPEC governments, thanks to their earlier successes, had financial reserves of a year or more.[56] Indeed, the production cutbacks in connection with the Arab embargo of 1973 showed that—within elasticity limits as yet unexplored—the less oil OPEC produced, the more it would receive in payment.

Among all OPEC countries, Venezuela has gone furthest in actually nationalizing, as of the beginning of 1976, all its oil production, including the necessary new investments for development of new reserves as production from old fields declines. Other governments, notably Algeria and Iraq, operate their fields with some help from foreign technicians, but still may be expected to rely on foreign companies for additional development. Elsewhere the international oil companies, under one kind of arrangement or another, still operate most of the production facilities, serving, in effect, as technical contractors. Countries such as Iran and Kuwait may be expected to move increasingly toward operation by their own national companies. Less populous, and hence talent-starved, countries such as Libya and Abu Dhabi, may have to continue to rely on foreign companies well into the future. Yet even in countries such as Iran, Saudi Arabia, and Kuwait ambitious plans for industrial development, both in the petrochemical and other sectors, are sure to make competing demands on the same growing pool of talent.[57]

In the countries where governments own the production concessions but international companies continue to operate them,[58] the companies receive, in effect, two sorts of compensation: access to large amounts of crude with which to supply their world-wide downstream operations and a discount (currently at just over 2%) [59] on that crude. Countries that wish to induce companies to undertake major prospecting and development programs typically offer additional benefits.[60] As noted earlier, the companies these days are concerned less about compensation for past investments long recouped than about future access to supplies of crude petroleum.

OPEC's continued reliance on the major multinational companies for the world-wide distribution of most of its oil implies one other essential function that is rarely recognized by observers of the international oil scene: the companies and the changing size of their markets in industrial countries continue to play a major role in adjusting levels of production among OPEC's member countries.

Under the pattern of concessions that developed in the 1920s, operations in most oil-producing countries were shared among two or more major companies. (The last country-wide concession held by a single company, that of BP in Iran, in 1954 was converted into a 15-company consortium.) Venezuela, Libya, Nigeria, and others, however, developed a different pattern of not conceding the entire country to a single company or consortium, but of parcelling out various smaller areas to different companies or groups. For their part, the seven major companies preferred to spread their economic and political risks over a number of countries. The net result is a doubly interlocking pattern whereby all seven major companies operate in the Middle East, six of them in Venezuela, five in Libya, four in Nigeria, and four in Indonesia, and each of the seven companies serves all three of OPEC's major markets: Europe, Japan, and the United States.

OPEC countries do some of the job of setting amounts of production themselves. The Arab countries during the embargo decreed successive cutbacks that the companies were required to follow. Some Arab and non-Arab countries, notably Venezuela, Kuwait, and Qatar, have set production limits for purposes of conservation. (Saudi Arabia, Abu Dhabi, and Libya also have decreed maximum limits, but production has not come up to these for some time.) [61] This still leaves a wide margin for variation of actual levels of production.

For example, early in 1975, OPEC's aggregate production capacity was estimated at 38.1 million barrels a day, and government limitations reduced this total to 34.05 mb/d. But world demand for OPEC oil in the first quarter of 1975 was down to an average of 26.5 mb/d. Within its own share of this total (ranging from about 20% for Exxon down to about 5% for Mobil) each of the seven major companies, therefore, had to decide how much of its current demand to meet from its sources in, say, Nigeria, Iran, Saudi Arabia, or Venezuela. The companies have strong incentives to make these allocations solely by objective criteria of quality (gravity and sulfur content), tax-paid cost per barrel, and transportation cost to market.[62] In any case, it is the aggregate of these amounts of company "offtake" that determine the month-to-month production levels in any one country. As Anthony Sampson recently noted, "the oil companies really [are] doing the countries' rationing for them." [63]

In sum, the revolution of 1971-75 leaves OPEC governments in control of world prices, with the right (sporadically exercised) of limiting production, and in receipt of about 97% of the revenue from the production of crude. It leaves the companies in charge of many of the technical operations, of allocating production among countries, and of world-wide distribution and sales. And although the companies' profits on the production of crude have sharply declined, down to about 2%, they now handle a far more expensive product, so that even a smaller profit margin is likely to yield larger absolute amounts—and the profitability of their "downstream" operations (transporting, refining, retailing) is undiminished.[64] The steep increases in OPEC's revenue represent a financial transfer not from the companies but indirectly from the consumers. OPEC's future prospects therefore depend largely on the response of the consumer countries.

Notes

1. See Charles Issawi and Mohammed Yeganeh, *The Economics of Middle Eastern Oil* (New York: Praeger, 1962), p. 129 for 1950; and Petroleum Information Foundation, *Background Information,* nos. 8 and 16 (New York: PIF, 1970 and 1973) for figures since 1954. For these later years Issawi and Yeganeh have slightly divergent figures, indicating no absolute decline for Kuwait or Saudi Arabia, a slight one for Iran (1959), and a major one for Iraq (1956, 1957).

2. Issawi and Yeganeh, *Economics,* p. 61. These figures exclude Soviet bloc countries. Company names as of 1976. An eighth company, Compagnie Française des Pétroles, shared in the concessions in Iraq, Iran (after 1954), and some other countries. Cf. Table 9 in the appendix.

3. "Transfer prices from one to another division of the same corporate entity are simply bookkeeping notations to permit the corporation to minimize its tax bill," as M. A. Adelman succinctly puts it in *The World Petroleum Market* (Baltimore: Johns Hopkins University Press, 1972), p. 161 (hereafter cited as Adelman, *WPM*). This was true for the prices on which the royalties to Middle Eastern governments were calculated until the mid 1950s. With the advent of the fifty-fifty agreements, the governments received the option of taking some of their royalty in kind and thus, by selling it in the open market, to test the realism of the transfer prices charged by the integrated companies. This latter option was used sparingly, however. For example, the Saudi Arabian government from 1963 through 1971 lifted only 2.5 million barrels out of Aramco's total production for those nine years of 7.7 billion barrels—or less than .003%. See *Multinational Corporations and United States Foreign Policy,* United States Senate, Committee on Foreign Relations, Subcommittee on Multinational Corporations, *Hearings* (hereinafter cited as *Church Committee Hearings*), vol. 8 (Washington: GPO, 1975), p. 467.

4. Prices per barrel of Saudi Arabian light (34.0° to 34.09°), ex Ras Tanura were $2.08 (September 8, 1958), $1.90 (February 13, 1959), $1.80 (August 9, 1960), and $2.18 (February 15, 1971). See OPEC, Statistical Unit, *Annual Statistical Bulletin 1973* (Vienna: OPEC, 1974), p. 118. On the price cuts of 1957 and the following years, see Adelman, *WPM,* pp. 160-91; Anthony Sampson, *The Seven Sisters: The Great Oil Companies and the World They Made* (London: Hodder and Stoughton; New York: Viking, 1975), pp. 150-58.

5. Benjamin Shwadran, *The Middle East, Oil, and the Great Powers,* 3rd ed. (New York: Wiley, 1973), pp. 13-47; annual production and revenues are listed on pp. 132f. Cf. Raymond F. Mikesell and Hollis Chenery, *Arabian Oil: America's Stake in the Middle East* (Chapel Hill: University of North Carolina Press, 1949), p. 181.

6. Zuhayr Mikdashi, *The Community of Oil Exporting Countries* (Ithaca: Cornell University Press, 1972), pp. 22-23; *Organization of the Petroleum Exporting Countries* (Vienna: OPEC, 1966), pp. 4-11. The Conference was attended by observers from Venezuela and Iran.

7. Iranian production, 1913-50, totalled 2.4 billion barrels (Issawi and Yeganeh, *Economics,* p. 183). In 1951-54, 59.2 billion barrels were added to Middle Eastern reserves. See *BP Statistical Review of the World Petroleum Industry, 1960.*

8. Shwadran, *op. cit.,* pp. 97, 103. Cf. Sampson, *Seven Sisters,* pp. 119ff.; the Royal Air Force was used to force tankers with Iranian oil into the harbor of the British Crown Colony of Aden, where the cargoes were impounded.

9. Shwadran, *op. cit.,* p. 165; and Ashraf Lutfi, *OPEC Oil* (Beirut: Middle East Research and Publishing Center, 1968), pp. 59-64. Lutfi was OPEC's secretary general in 1965-66. Cf. Sampson, *Seven Sisters,* p. 165.

10. See Fuad Rouhani, *A History of OPEC* (New York: Praeger, 1971), for Resolution IV.32 (pp. 195f.) and IV.33 (p. 221); IV.34 (p. 245) urged the elimination of the contribution of about two cents per barrel toward marketing expenses that the companies had been charging against the government's account. Rouhani was OPEC's first secretary general.

11. On the decline of prices for products and for crude in 1957-62,

see Sam Schurr and Paul Homan, *Middle Eastern Oil and the Western World* (Baltimore: American Elsevier, 1971), pp. 121-22; Michael Tanzer, *The Political Economy of International Oil and the Underdeveloped Countries* (Boston: Beacon, 1969), p. 73; and Neil H. Jacoby, *Multinational Oil* (New York: Macmillan, 1973), pp. 232-42. According to Jacoby (p. 239), the average price of petroleum products imported into Belgium declined from $5.11 per barrel in 1957 to $2.91 in 1962, after which it rose gradually to $3.27 in 1970. See also the detailed price statistics in Ente Nazionale Idrocarbui, *Energia ed Idrocarburi: Sommario Statistico, 1955-70,* 2 vols. (Rome, n.d.).

12. Rouhani, *History of OPEC,* pp. 220, 229-32. The royalty at this time was 16⅔% in Venezuela and 12.5% elsewhere. To make the transition to this new system less abrupt, the governments granted the companies a set of diminishing discounts from the posted price—8.5% for 1964, 7.5% for 1975, and 6.5% for 1966.

13. Petroleum Information Foundation, *Background Information,* no. 8, p. 3.

14. Edith Penrose, "The Development of Crisis," in Raymond Vernon, ed., *The Oil Crisis: In Perspective (Daedalus,* vol. 104, no. 4, Fall 1975), p. 40. A notable exception was Professor Penrose herself, who drew attention to the "Declaratory Statement" in a paper on "The Role of OPEC in Changing Circumstances" delivered at an OPEC seminar held in Vienna in July 1969, first published in *Middle East Economic Survey,* vol. 13, no. 49, October 3, 1969, and reprinted in her collection, *The Growth of Firms, Middle East Oil and Other Essays* (London: Cass, 1971), pp. 236-44, esp. p. 242. The same essay comments on what she describes as Sheikh Yamani's proposal for "an alliance between companies and governments" (p. 239) presented in his speech cited in the following notes.

15. Yamani's presentation and responses to questions are sum-

marized in detail in *Middle East Economic Survey,* vol. 11, no. 32, June 7, 1968.

16. The full text of the Declaratory Statement appears in the appendix.

17. World exports (f.o.b. value in 1963 prices) increased from $53.2 billion in 1948 to $285.2 billion in 1972. Global energy consumption rose from 2.52 billion tons coal equivalent in 1951 to 7.09 billion in 1971. From 1950 to 1972 the world domestic product increased by 225%. *United Nations Statistical Yearbook 1973* (New York: United Nations 1974), p. 10, 55, and *idem, 1965* (New York: United Nations 1966), p. 26. On energy consumption, cf. Sam H. Schurr and Paul T. Homan, *Middle Eastern Oil and the Western World, Prospects and Problems* (New York: Elsevier, 1971), p. 173.

18. By comparison, the production of a ton of coal, equivalent in heating power to about 4.9 barrels of crude oil, cost $6.61 in the United States in 1959, and as much as $16 to $19 in Europe. *Trends in Electric Utility Industry Experience 1946-1958* (Washington, D.C.: National Coal Association, 1960), p. 96, has the U.S. data from 1952. *Steam-Electric Plant Factors 1959* (Washington, D.C.: National Coal Association, July 1960), p. 19, indicates that in the United States during 1959 the cost of coal per ton "as consumed" was $6.61 and the cost of a barrel of oil "as consumed" was $2.09. United Nations, Secretariat, Statistical Office, *Commodity Trade Statistics,* Series D, Vol. IX, Nos. 1 and 2, January-June 1959 (New York: United Nations, 1960), pp. 138, 309-10, has the export price of coal in Western Europe. For estimates of petroleum production costs in each OPEC country for 1974, see Table 4 in the appendix.

19. Calculated from *BP Statistical Review of the World Oil Industry 1960.* Cf. Table 1 in appendix.

20. Development costs (1953-1962) in the Middle East averaged 11 to 15 cents a barrel, and 39 cents per barrel in Venezuela; see Paul

Bradley, *The Economics of Crude Petroleum Production* (Amsterdam: North Holland Publishing Co., 1967), p. 102. Libyan operating costs per barrel decreased from 9.5 cents (1964) to 7.6 cents (1965), and Iranian operating costs from 8.09 cents (1962) to 4.81 cents (1969). Adelman, *WPM,* pp. 288-89. For 1974 estimates, see appendix, Table 4.

21. No less an authority than Professor Adelman *(WPM,* p. 256) has insisted that "Nobody knows, even approximately, how high the OPEC nations can raise taxes, and thereby prices, before sales drop off so rapidly as to reduce their total take." He quite presciently foresaw in 1972 that "It is a much higher prize than what the producing countries have yet achieved"; and suggested that OPEC countries may well succeed in forcing a transfer of the monies so far levied as excise taxes in Europe to themselves. All this, of course, is only part of Adelman's complex and dialectic argument, which concludes (p. 261) that "in the end [sic] the United States and the other consuming nations can have the price of oil they want."

The point in the text is not how much OPEC alone can extract, but how much companies, producing governments, *and* consuming governments can extract from the consumer without diminishing the aggregate take of the three. What the consumer paid in 1975 in Europe, which is OPEC's largest customer, is approximately and in very round figures $22 per barrel, of which $10 is taxpaid cost at the Persian Gulf, $1.50 the cost of transport and refining, and $10 local taxes (taking a rough weighted average of the tax rates listed in Table 8, in the appendix). Only when the retail price to the consumer went to this level from the earlier price of roughly $15 per barrel was there a beginning of consumer resistance. But a price rise of about 50% brought a decline in volume of less than 10%. Hence our (carefully hedged) statement in the text.

22. Stephen Hemsley Longrigg, *Oil in the Middle East* (New York: Oxford University Press, 1954), pp. 67ff.

23. Figures for 1967 from D. A. Rustow, *Middle Eastern Political*

Systems (Englewood Cliffs, N.J.: Prentice-Hall, 1971), p. 17. The remaining 2% represented Gulbenkian (the legendary "Mr. Five Percent" of the Middle East petroleum industry, whose proverbial share, however, applied only in Iraq and some smaller areas) and the National Iranian Oil Company. In 1947 the ratios had been American 42%, British 30%, British-Dutch 22%, and French 6%; see Mikesell and Chenery, *Arabian Oil,* p. 177.

24. J. E. Hartshorn, *Politics and World Oil Economics* (New York: Praeger, 1962), pp. 175ff., gives a detailed account of the relevant tax provisions; the quotation is from p. 199. The British edition of this work is entitled *Oil Companies and Governments* (London: Faber, 1962). On the Venezuelan precedent for the fifty-fifty arrangement, see William G. Harris, "The Impact of the Petroleum Export Industry on the Pattern of Venezuelan Economic Development," in Raymond F. Mikesell, ed., *Foreign Investment in the Petroleum and Mineral Industries* (Baltimore: Johns Hopkins University Press, 1971), pp. 129-59, esp. p. 107. The largest foreign oil company, an Exxon affiliate, agreed to pay the 50% tax retroactively to 1946.

25. Robert B. Krueger, *The United States and International Oil* (New York: Praeger, 1975), pp. 15, 71. This study was prepared for the Federal Energy Administration.

26. Krueger (p. 52) states that Aramco's tax payments in 1950 were $50 million to the U.S. and $56 million to Saudi Arabia, and in 1951 $6 million and $110 million, respectively—a net increase of only 9.4%, even though production increased by 40%. According to "A Report of the Staff of the Joint Committee on Internal Revenue Taxation" of the U.S. Congress reprinted in *Church Committee Hearings,* vol. 8, pp. 350-78, at p. 357, the fifty-fifty arrangement in Saudi Arabia "resulted in credits increasing from $46 million in 1950 to almost $177 million in 1955. In 1950, 1951, and 1952 the net tax paid by Aramco to the United States amounted to less than $1 million. In each of the years since that time the credit has completely offset the United States tax."

27. 1948 data for the five countries that later founded OPEC are calculated from Issawi and Yeganeh, *Economics,* p. 188f; for later figures, see Table 2 in the appendix.

28. On the basing point practice, see Issawi and Yeganeh, *Economics,* pp. 64-70, who also calculate (pp. 42, 121) that total investments by the major oil companies in the Middle East from 1926 to 1960 amounted to $2,375 million (at historical cost), and that their net income from Middle Eastern operations from 1948 to 1960 averaged $1,352 million per year.

29. S. David Freeman, *Energy: The New Era* (New York: Walker, 1974), p. 117.

30. *Pétrole 71* (Paris, 1972), p. 289, details the tax rates imposed on gasoline by European governments in 1950, 1960, 1971, and 1972. ENI (cited in note 11, above) pp. 100ff, gives similar figures for the last months of each year 1960, and 1965-71. The range began at 34% to 73% in 1950 and went to 58% to 78% in 1972, with France and Italy having the highest, and Germany and the United Kingdom doubling theirs in these two decades, Cf. Table 8 in the appendix.

31. See Paul Frankel, *Mattei: Oil and Power Politics* (London: Faber; New York: Praeger, 1966).

32. The Getty Oil Company took all of the Saudi Arabian half-share in the Neutral Zone and ten non-majors formed the American Independent Oil Company to exploit the entire Kuwaiti share. For further information see Roy Lebkicher et al., *ARAMCO Handbook* (New York: Arabian American Oil Company, 1960), pp. 115f. By 1971, six majors accounted for 40% of Libyan crude oil production, with 20 independents accounting for the remaining 60%. For shares of Libyan production, see OPEC Statistical Unit, *Annual Statistical Bulletin 1971* (Vienna: OPEC, June 1972), p. 41, and for owners of the various concessions, see *Arab Oil and Gas* (March 1974), pp. 32-34; see also Table 9 in the appendix.

[40]

33. This applies to exports from OPEC countries. OPEC Statistical Unit, *Annual Statistical Bulletin 1971,* p. 40. Jacoby, *Multinational Oil,* p. 177, calculates the share of the seven majors in oil production outside the United States at 87.1% in 1953 and 70.9% in 1972. Cf. Table 9 in appendix.
34. See notes 4 and 11, above, for price data, and Table 1, appendix, for the magnitude of market expansion as measured by the increase in world imports.

35. For world and Western European imports, see *BP Statistical Review 1960,* p. 21, and *idem, 1970,* p. 21. For revenue figures for 1950, see Issawi and Yeganeh, *Economics,* pp. 121; and for 1970 see Petrolem Information Foundation, *Background Information,* no. 8, pp. 2f. Cf. Tables 1 and 2 in the appendix.

36. The OPEC share of world crude oil exports was 83% in 1960 and 82% in 1970; see Table 1 in the appendix. Oman with a yearly production of about 0.3 mb/d since 1970, is the largest Third World exporter to have remained outside OPEC. It has, however, matched step for step the tax increases of other Persian Gulf producers.

37. M. A. Adelman, "Is the Oil Shortage Real?" *Foreign Policy,* no. 9 (Winter 1972-73), pp. 80f. For a rebuttal by the State Department's chief oil expert in those years, see James E. Akins, "The Oil Crisis: This Time the Wolf Is Here," *Foreign Affairs,* vol. 51, no. 3 (April 1973), pp. 462-90, esp. pp. 472f.

38. Sir Eric Drake, quoted by Adelman in "Is the Oil Shortage Real?" pp. 70 and 78.

39. Adelman, *WPM,* p. 224.

40. Production in September 1970 was 800,000 barrels per day less than in April *(Petroleum Press Service,* October 1970, p. 358; hereafter cited as *PPS);* tanker rates moved from 95 to 261 Worldscale *(PPS,* September 1970, p. 353), a level not attained since the 1967 Arab-Israeli War and one that was to rise steadily

further until November 1970. European demand in January-June 1970 was up 12.4%, American up 5.5%, and Japanese up 22.9% over January-June 1969 *(PPS,* November 70, pp. 402-3.) In Syria, a bulldozer had rammed the TAP-line and the government took its time about allowing repairs. On Qaddafi's tactic in dividing the companies, see Sampson, *Seven Sisters,* p. 212; Occidental, on which the pressure was turned first, asked Exxon for a promise of crude at near cost if it held out against Qaddafi's demand and had to shut down. Exxon promised to furnish the crude only at normal third party prices, and Occidental gave in.

41. *PPS,* March 1971, p. 82.

42. Peregrine Fellowes, "Living Dangerously—Can Europe Afford to Rely on Middle East Oil?", *New Middle East,* October 1971, p. 26.

43. Resolutions XXI.120, 122, 124.

44. Resolution XXI.131.

45. Resolution XXVII.145.

46. Resolution XXVIII.146.

47. The rapid developments summarized here can be followed at leisure in the trade press, e.g., *Petroleum Press Service* (London; in 1974 renamed *The Petroleum Economist,* cited as *TPE), Petroleum Intelligence Weekly* (New York; cited as *PIW),* and *OPEC Annual Review and Record 1973* (Vienna: OPEC, 1974), as well as the *New York Times.* Good summaries will be found in *Keesing's Contemporary Archives* (London). For the background of certain crucial episodes, see, e.g., Edward Sheehan, "Colonel Qadhafi—Libya's Mystical Revolutionary," *New York Times Magazine,* February 26, 1972, and Joseph Kraft, "Letter from OPEC," *The New Yorker,* January 20, 1975. For an excellent interim assessment see John C. Campbell and Helen Caruso, *The West and the Middle East* (New

York: Council on Foreign Relations, 1972), esp. pp. 47-52. On the details of the Tehran agreement, see e.g., *New York Times,* February 15, 1971, and *Neue Zürcher Zeitung,* February 17, 1971. The full text appears in OPEC, Information Department, *Documents of the International Petroleum Industry 1971* (Vienna: OPEC, 1973), pp. 391-96.

48. *OPEC Annual Review and Record 1973,* pp. 8, 47; *TPE,* January 1974, p. 9. The corresponding government revenues were $1.77, $3.05, and $7.01 per barrel. Changes in 1974 increased the latter figure retroactively to $9.27 as of January 1. See Table 3 in the appendix.

49. *New York Times,* November 24, 1973; for monthly production figures, see Table 5 in the appendix. For an excellent analysis of the handling of the embargo by producer countries and companies, see Robert B. Stobaugh, "The Oil Companies in the Crisis," in Vernon, ed., *The Oil Crisis (Daedalus),* pp. 180-202. See also Federal Energy Administration, *U.S. Oil Companies and the Arab Oil Embargo: the International Allocation of Constricted Supplies*, U. S. Senate, Committee on Foreign Relations, Subcommittee on Multinational Corporations, Committee Print (Washington: GPO, 1975).

50. This statement is based on the following calculations, reproduced here from Rustow, "Who Won the Yom Kippur and Oil Wars?" *Foreign Policy,* no. 17 (Winter 1974-75), p. 168:

	United States	European Community	Japan
1. Energy consumed per capita *	8.2	3.6	3.1
2. Per cent of energy from oil	47.2	59.5	76.4
3. Per cent of oil imported	36.9	98.7	99.7
4. Per cent of energy from imported oil (2 x 3)	17.4	58.7	76.2
5. Per cent of oil imported from Arab sources **	29.4	66.0	42.0
6. Per cent of energy from Arab sources (4 x 5)	5.1	38.8	32.0
7. Energy per capita from sources other than oil imports (1-[1 x 4]) *	6.8	1.5	0.7

51. *New York Times,* December 29, 1973, January 28, 1975, January 30, 1975. This last issue reported that export price of Canadian oil would soon rise from $11.70 per barrel to $12.10.

52. The latter, in turn, represented production cost, a modest profit, plus a weighted average of payments to the government (income tax and royalty on the company's oil and the buy-back price for oil purchased from the government's share). Cf. Table 3 in the appendix.

53. *New York Times,* January 10, 1974, March 18, 1974, September 14, 1974, September 17, 1974; *TPE,* July 1974, p. 251; November 1974, pp. 413-15; December 1974, pp. 454-55; *Middle East Economic Survey,* December 6, 1974, p. 1.

54. *TPE,* May 1974, p. 175; August 1974, p. 296.

55. *TPE,* October 1974, p. 375; *New York Times,* November 30, 1974, December 14, 1974; December 23, 1974, December 24, 1974; *Wall Street Journal,* December 6, 1974.

56. In 1970, the foreign exchange reserves of Libya substantially exceeded the value of its annual imports. See below, p. 96 and Table 6 in the appendix.

57. The participation agreements concluded in 1972 by Saudi Arabia and other Arab Gulf countries and the Venezuelan Hydrocarbon Reversion Law of 1971 envisaged a full takeover by the early 1980's, and the Iranian agreement with the consortium of 1973 guaranteed the consortium members' access to their normal quotas of

---

* Tons oil equivalent per year.

** Before the October 1973 embargo.

Sources: The calculations (for 1973, except as noted) are from *BP Statistical Review 1973* (lines 1, 2, 3); U. N., *Monthly Bulletin of Statistics,* August 1974 (line 1, population); *World Oil,* December 1973 (line 3, Japan); OECD, *Oil Statistics 1972,* pp. 26f. (line 5, Japan, 1972); *Petroleum Press Service,* November 1973 (line 5, Europe); U.S. Bureau of Mines, *Mineral Industry Surveys* (line 5, United States).

oil through 1995. Developments since then indicate that the nationalization schedule implicit in these dates has been considerably foreshortened.

58. The Iranian situation indicates how subtle the legalisms of nationalization can get. Mossadegh nationalized the oil in 1951, but in 1954 the Consortium obtained the right to operate the production of this nationalized oil. In 1975 the Shah nationalized the production as well, but the Consortium companies stayed on as contractors for the National Iranian Oil Company, and were guaranteed their normal share of production for twenty years.

59. E.g., from January to September 1975, Arabian light ex Ras Tanura was sold, whether by the Saudis directly or by Aramco, at $10.46 a barrel. For Aramco's equity portion, this total divided as follows: production cost 10 cents (0.96%), payments to Saudi government $10.12 (96.75%), company net revenue from production 24 cents (2.29%). These figures, based on trade press reports, differ slightly from those given in Table 4 in the appendix, but indicate the basic orders of magnitude.

60. Thus, Algeria in January 1975 announced a price of $12.50 with a $0.20 per barrel discount for countries willing to make exploration investments *(New York Times,* January 5, 1975).

61. The following production limits were put into effect in 1974 and 1975: Saudi Arabia 8.5 mb/d (March 1974); Kuwait 2 mb/d (January 1975); Abu Dhabi 1.6 mb/d (March 1974), 1.3 mb/d (September 1974), 1.5 mb/d (January 1975); Qatar, 0.5 mb/d (January 1974); Libya 2.3 mb/d (early 1974). Venezuela in early 1973 limited the production of oil to such amounts as can be produced without flaring associated gas; this has meant a gradual lowering of production from 3.4 mb/d in 1973 to 2 mb/d by late 1975. There are no limitations for the Saudi-Kuwaiti Neutral Zone or for Kuwait offshore. See also Tables 5 and 7 in the appendix.

62. Only when a country overprices its crude in relation to these objective factors, as Abu Dhabi and Ecuador did in 1974-75, will the companies feel justified in cutting production disproportionately to other countries. In both countries, this had the effect of appropriate price reductions—which indicates that, whereas the major Persian Gulf producers and Venezuela set the basic price level, the companies play an important role in keeping the lesser, outlying OPEC producers in line.

63. *Seven Sisters*, p. 301.

64. For example, average oil company payments to governments in 1972-73 rose from $1.55 to $2.12 (see Table 2); whereas their net earnings per barrel jumped from $0.283 to $0.698, and total net earnings from $2.0 billion to $5.2 billion. (The last two pairs of figures refer only to the Eastern Hemisphere operations of the seven major oil companies; see First National City Bank, *Energy Memo*, January 1975. No comparable figures for 1974 were available at the time of writing.) Cf. Table 2 in the appendix.

[TWO]

Containment and Expansion

Fears and Hopes

The world's industrial economy had rarely sustained as sudden a shock as the petroleum crisis of 1973-74. No wonder that even seasoned observers were perplexed, their moods ranging from panicky to serene and their advice running the gamut from confrontation to accommodation.

How long will the cartel last? What will be the future price of oil? What financial surpluses will the oil countries amass? And what will be the effect on capitalist economies throughout the world? These were the questions that political and economic analysts anxiously pondered.

The sanguine answers in the United States and other consuming countries were that OPEC would break up; that the price would come down, if not in 1976 or 1977, then in 1980 or soon after; and that everything would be back to normal in due course. "The cartel's weak spot," as Professor

Adelman has insisted, "is excess capacity." [1] An inescapable glut was building up in the world oil market. Higher prices and government conservation measures were depressing demand. New discoveries in Alaska, the North Sea, and elsewhere were increasing the supply. Sooner or later OPEC would have to face the long-evaded need for imposing strict production quotas. But any such attempt would sharpen latent economic and political differences among the members. Even if OPEC managed to draw up a production schedule, members were almost certain to shave prices through a variety of hidden discounts in a competitive effort to expand their shares of a shrinking market. And once started, such a downward price spiral would acquire inexorable momentum. In sum, the optimists among consumer country experts held with Professor Adelman that OPEC would prove no exception to the rule that "Every cartel has in time been destroyed by one, then some members chiselling and cheating . . ." [2] Or as William E. Simon, first U. S. Energy Administrator and later Secretary of the Treasury, kept repeating, the question is not whether the price of oil will come down but when.[3]

The part that collective policies of consumer governments might play in weakening OPEC was carefully spelled out by Secretary Kissinger in a speech to the International Energy Agency in May 1975. Even in "the short term," "rigorous conservation and development of alternatives" will add to "surplus capacity"; cause "the producers' market . . . to shrink, first relatively and then in absolute terms"; and put "individual producers—especially those with ambitious development, defense or other spending programs . . . under pressure to increase sales or, at least, to refuse further production cuts." Thus "the cartel will have lost [its] . . . arbitrary control over prices," and the aim of restoring "balance in the international energy market" will have been achieved.[4]

Heedless of such hopeful speculations, OPEC continued to increase its tax take, from the $7 level decreed at Tehran in December 1973 to a level of $9.27 applied in mid-1974 retroactively to January 1, and to $9.98 on January 1, 1975, and $11.00 in October 1975. Despite many reports of wrangles behind the scenes of OPEC meetings, there was no sign of any open disputes. As a *Fortune* correspondent exclaimed, in some surprise, "Practically the only action they ever fully agree on is raising the price of crude oil." [5] More cautious observers arrived at a new appreciation of OPEC's strength. The cartel, they began to consider, might last for a decade or more. The limiting factor, on this line of analysis, would be the development of alternatives to petroleum in large quantities. But any expansion of nuclear energy and any full-scale development of coal gasification or liquefaction (not to mention costlier or less advanced technologies, such as shale, solar, geothermal, or tidal energy) would take at least until 1985, if not until the end of the century. Even the optimistic "Project Independence" report, compiled by the Federal Energy Administration, did not envisage even temporary freedom from imports before the mid-1980s.[6] For other industrial regions, such as continental Europe and Japan, there was no possibility of such independence. An OECD report of late 1974 estimated that by 1985 imported oil would still supply at least 41% of the total energy requirements of Western Europe and as much as 80% of Japan's.[7] The less sanguine observers, therefore, concluded that OPEC and its price structure would endure into the late 1980s or beyond.

Assessments such as these were translated into more precise forecasts in the research departments of international agencies, large banks, and consulting firms. The OECD report just mentioned considers three price levels, a "base case" of pre-October 1973 prices (that is, about $2.50 per barrel), and two other cases of $6 and $9 per barrel in 1972 dollars—the latter corresponding roughly to $7.20 and $10.80 a barrel in end-of-1974 dollars.[8] The First National City Bank detailed a "central scenario" by which the price of oil, in current dollars, would rise slightly until 1976 and then decline to $9.10 by 1980. Irving Trust posed a twofold alternative: a gradual rise to $15.72 by 1980 ("Case A"), and a price break that would bring the price down a dollar per year, from $12 in 1976 to $7 in 1980 ("Case B"). Morgan Guaranty's figure for 1980 was $13.72 per barrel.[9] Walter J. Levy, in a careful examination, considered a "base case," with a barrel of oil costing $14.65 in current dollars in 1980, and two variations: one with the price remaining at $10 in current dollars, hence declining with world inflation, and the other with the price going up gradually to the level of Irving Trust's "Case A." [10]

These forecasts, all published within a six-month period, diverge widely, the 1980 oil prices they envisage ranging from $7 to $15.72 a barrel. Still, they consider only a very narrow span of the imaginable alternatives—or "futuribles" as Bertrand de Jouvenel would have called them. Specifically they suggest: (1) that the price of oil will remain at current levels with periodic upward adjustments for world inflation; or (2) that the price will remain at current levels in nominal terms, gradually eroded by inflation; or (3) that a gradual increase will be followed by a price break that will restore prices approximately to the real level they had reached at the time of the 1973 oil embargo.[11] None of the forecasters, that is to say, dwells on two other possibilities: (4) that the price of oil will return to the $2 a barrel level of the 1960s, constant in nominal (and hence declining in real) terms; or (5) that OPEC will continue pursuing its advantage by raising prices by more than the global rate of inflation, thus bringing them nearer to the monopolist's optimum (of $20-plus?).

[50]

Estimates of the surplus funds that the OPEC countries may be expected to accumulate over the years depend on several assumptions in addition to the price of oil. How much oil will OPEC export at a given price level? How much will OPEC countries import in goods and services? (And this in turn depends on further questions such as the capacity of ports and other transport facilities, and the capacity of economic planners to draw up and execute ambitious plans.) Finally, what returns will OPEC countries earn on their investments of surpluses of prior years? Even small divergences on one of these points build up over the years to large differences. No wonder, therefore, that while the estimates of the price of oil for 1980 cover only a 2:1 range, estimates of OPEC's financial surpluses cover a range of approximately 25:1—a sobering reflection on the hazards of economic forecasting. The specific predictions are shown in the following table:

OPEC Cumulative Financial Surplus, End of 1980
As Variously Estimated (Billions of Current Dollars)

Source	Date of Publication	Estimated OPEC 1980 Surplus
Irving Trust (Case B)	March 1975	$ 22 billion
Morgan Guaranty	January 1975	179
First National City Bank	May 1975	189
Irving Trust (Case A)	March 1975	248
OECD ($6 Case) *	December 1974	393 to 510
K. Farmanfarmaian et al.	*Foreign Affairs*, January 1975	400 to 450
Walter J. Levy	June 1975	449
H. B. Chenery *	*Foreign Affairs*, January 1975	471
G. Pollack	*Ibid.*, April 1974	500
OECD ($9 Case) *	December 1974	510

* The OECD and Chenery estimates are in 1974 dollars and have been multiplied by 1.57 to convert them to 1980 dollars; cf. Levy, *Future OPEC Accumulation of Oil Money*, Table 4-A.

Although there is little agreement on the overall size of the OPEC surplus, there seems to be a general consensus that the bulk of these surpluses will accumulate in the Arab states of the Persian Gulf, notably in Saudi Arabia and Kuwait. For example, the First National City Bank study suggests that, according to its "central scenario," 48% of the total 1980 surplus will accumulate in Saudi Arabia and 25% in Kuwait. (According to the same estimate, Nigeria, Venezuela, Qatar, and Iraq will each hold surpluses amounting to 13% to 3% of the total; Abu Dhabi, Libya, and Ecuador will hold relatively insignificant positive balances; Iran will go into deficit after 1979; and Indonesia and Algeria will be in constant deficit for the entire 1975-80 period.)

Estimates of the impact of the oil crisis on the global economy ranged from fears of catastrophe to assurances of business as usual. The alarmist view was expressed early by Secretary of State Henry A. Kissinger in his address on February 11, 1974, to the Washington Energy Conference. Unless immediate steps were taken to overcome the oil crisis, the Secretary warned, the world would be threatened "with a vicious cycle of competition, autarchy, rivalry, and depression such as led to the collapse of the world order in the thirties." [12] The new tax take, then still fixed at just over $7, would set off a shock wave of demestic recession and umployment, and thus create intolerable strains on democratic political systems.

More specific apprehensions centered on the international financial system. Banks dealing in the so-called Eurocurrency market soon were caught between the short-term deposits of the *nouveaux riches* OPEC countries and the medium- or long-term borrowing needs of their customers. The failure of one or another major bank might soon set off a chain effect that would sweep down the entire house of cards of international banking and currency. Temporary foreign trade restrictions erected by Italy and Denmark and the failure in mid-1974 of the Bankhaus Herstatt of Cologne and the Franklin National Bank of New York seemed to indicate that these were no idle fears. And as late as September 23, 1974, President Ford confessed in a public address that "It is difficult to discuss the energy problem without lapsing into doomsday language." [13]

A proponent of a more sanguine view—on behalf of producers and consumers alike—was Hollis B. Chenery, a Vice President of the International Bank for Reconstruction and Development. The current transfer of funds from the industrial to the oil countries, Chenery reflected, was no larger (in proportion to the size of the current global economy) than had been the monetary transfers of the Marshall Plan. Therefore, he concluded, "it is difficult to argue on economic grounds that the world economy cannot sustain expected flows of the required magnitude, or that OECD countries need to suffer heavily in the process." [14]

Into this continuing debate about the dimensions and repercussions of the oil crisis an additional note of alarm was injected by C. Fred Bergsten, a Brookings Institution economist, who anticipated that OPEC would have a palpable demonstration effect on Third World countries producing other raw materials. The industrial countries would be faced with a whole series of cartels in a wide range of materials from bananas and bauxite to zinc. And indeed Jamaica's action in unilaterally quadrupling its levy on bauxite and bustling meetings of representatives of countries producing copper, coffee, tin, zinc, and other vegetable and mineral materials lent substance to such warnings.[15] Among the OPEC countries themselves the government of Algeria became the most vocal advocate for applying the OPEC tactic to other raw materials.

If there was little agreement about what would happen, there was less about what was to be done. Some observers hoped, by some kind of confrontation of consumers with producers of petroleum, to restore the situation to its pre-October 1973 normalcy. Others accepted the new realities, like it or not, but sought to improve on them through some bold, positive initiative. Still others, shunning grand designs of any sort, sought to cope from day to day and month to month with the dangers and opportunities of the situation.

The confrontation imagined by the drastic activists ranged from military invasion to economic retaliation to so many rounds of glowering at the conference table. At a weekend retreat of presidential advisers at Camp David, Maryland, that laboriously debated the impact of the oil crisis, one of the more waggish participants is said to have passed around a note reading "Let's consider the cheaper option: invasion." More deliberately, Secretary Kissinger adumbrated this possibility in an interview in late December 1974, notable both for its diplomatic innuendo and its syntactic bravado: "I am not saying," the Secretary protested, "that there's no circumstance where we would not use force. But it is one thing to use it in a dispute over price, it's another where there's some actual strangulation of the industrial world . . . The use of force would be considered only in the gravest emergency." [16] If mere disputes over price were excluded, the "grave emergency" constituting "some actual strangulation" presumably would be a renewed embargo and production cutback that surpassed those of 1973 in effectiveness. Possible military scenarios were spelled out in rather lurid tones in *Harper's* in March 1975 and more soberly by *The Economist.*[17] The underlying strategic and diplomatic calculations were carefully examined by Professor Robert W. Tucker in two articles in *Commentary*. Unless we are willing at least to discuss military action in defense of what is widely acknowledged to be a vital interest of the United States and all its industrial allies, Tucker concluded, we surely are paying for a vastly overpriced military establishment.[18]

Others were hoping to counteract OPEC through economic or political rather than military means. Some pointed out that the oil countries depended on the United States and other Western countries for much of their food and nearly all their machinery, and specifically for their arms and their oil drilling equipment. If they could embargo our oil, why could we not embargo their food and machinery? [19]

Some believed that OPEC would never have acquired its stranglehold except for the willing cooperation of the major oil companies and for the advice of the Department of State in 1970-71 urging acceptance of the oil countries' demands.[20] This suggested, as a remedy, taking the companies out of the international oil trade by centralizing all U.S. oil imports in a single governmental buying agency that would ask for competitive sealed bids from producing countries or their middlemen. Eager for maximum shares in the large American market, suspicious of their fellow OPEC members, and assured of secrecy, the oil countries would thus have a potent motive for competitively cutting their prices.[21] Or perhaps OPEC's power could be broken if the companies were relieved of their function as "tax-collecting agency" for the producer countries—e.g., by cancelling the foreign tax credits that allow the companies to deduct a large part of their payments to OPEC from U.S. taxes on their foreign operations. Since the tax credit provisions, it is argued, encouraged the companies to yield to the oil governments' financial pressures in the first place, their cancellation would induce them to put up more resistance. Being forced, that is, to pay higher taxes at home, they might choose to stop being OPEC's tax collectors.[22]

Another school of thought advocated coping with the oil crisis by restricting consumption in the industrial countries and hence curtailing oil imports from OPEC. For example, France under President Giscard in 1975 limited annual oil imports to a sum in French francs about 10% below the previous year's imports—which had the obvious advantage of setting a limit to the additional foreign exchange drain imposed by the oil crisis. A rather different device was President Ford's imposition of a tax (first of $1 and then of $2) on each imported barrel of oil. This, it was hoped, would not only discourage imports and reduce consumption but also, by raising the price of oil within the United States, encourage oil companies to increase domestic production of petroleum and of various alternatives. Whatever the method of curtailing imports, many observers hoped that a substantial reduction of OPEC's world market would bring the organization that much faster to the point of having to allocate production, and hence of breaking up the cartel as members fought to enhance their shrinking market shares.

In sharp contrast to the advocates of some kind of confrontation with OPEC—polite or drastic, military, diplomatic, or economic—stood those observers who considered OPEC fairly immune for some years ahead and hence wished to overcome the crisis in a spirit of accommodation or positive cooperation with the producers. Even the leading advocates of collective action by the consumer countries, such as oil consultant Walter J. Levy and Secretary of State Kissinger, tended to emphasize that the purpose must be not confrontation with the producers but the elaboration of some set of mutual interests.[23]

Others groped for some more dramatically positive response that would improve relations with OPEC or with the Arab countries making up its majority. In this category were the diplomatic communiqués endorsing the Arab position on the conflict with Israel issued by the European Communities and by the Japanese government at the time of the oil embargo and the preparations for a "Euro-Arab dialogue" begun soon after.[24]

Still others, such as Dr. Chenery of the World Bank, conceived the problem more broadly as one of "Restructuring the World Economy," [25] and non-Western leaders, notably Boumedienne of Algeria and Luis Echeverría of Mexico, stepped up their efforts to create a "new international economic order." One concrete effort along those lines was the preparatory meeting for a tripartite conference of oil-consuming countries, oil-producing countries, and non-OPEC Third World countries held in Paris in April 1975 and the full-fledged Conference on International Economic Cooperation which opened, also in Paris, on December 16, 1975. At the preparatory meeting Algeria emerged as the champion for a new economic order involving stabilization of prices for all raw materials; the United States delegation, however, adamantly refused to extend the discussion from oil to other raw materials; and when the European representatives sided with the Americans, the conference adjourned in failure.[26] Somewhat surprisingly, this position was reversed almost immediately: the United States, it now appeared, would be ready to talk about other raw materials, provided these were discussed one by one in separate subcommittees.[27]

One of the most ingenious proposals for coordination first among oil-consuming countries and then between oil consumers and oil producers was based on the plan to set a long-range "floor price" for oil, which was elaborated in the State Department in the winter of 1974-75 and championed by

Assistant Secretary Thomas O. Enders at a meeting with other OECD governments in Paris in April 1975.[28] The floor price proposal was based on a subtle line of reasoning that ran essentially as follows: The price of oil will come down from its current excessively high levels only as the result of the development on a large scale of alternatives to petroleum. Such development will require the commitment of a major portion of the investment resources of the industrial world. Yet it could be thwarted, just as it came to fruition, by OPEC reducing the price of its oil to just below that of the alternatives; and in the face of such a risk, the necessary investments were not likely to be made in the first place. Therefore, in committing themselves to such an investment program, the major industrial countries—North America, Western Europe, Japan—should also promise each other not to let the price of energy, including any oil imports from OPEC, drop below levels required to amortize this investment. Such a long-range floor price will deprive high priced petroleum of any competitive advantage, and hence threaten OPEC with total loss of its markets once the alternatives are in full production. Therefore some advocates of the floor price plan thought that OPEC, rather than insisting on its cartel price now and facing complete collapse later, would accept a stable intermediate price, say $7 or $8—that is, the very floor price agreed upon among industrial countries.

Thus the debate continued between those who proposed some joint action to bring down the price of oil and those who envisaged some grand or minute scheme that would allow the world to live with it. The vacillations of U.S. policy, meanwhile, have contributed to delaying the development of new petroleum sources and of alternatives, and the periodic hints of confrontation have slowed the potential flow of OPEC investment into the United States, for both development and investment would require, above all, a stable and hospitable political climate.

Security of Supply and the Recycling of Funds

The crisis in the autumn of 1973 added four major items to the diplomatic agenda: the need to safeguard against a new oil embargo, the payments difficulties of countries faced with quintupled oil bills, the attendant strains in the international monetary system, and the demonstration effect that OPEC might have on Third World producers of other raw materials. Concern about the security of supply was, naturally, most acute among the major industrial oil-importing countries. The preoccupation with what became known as the "recycling of petrodollars" was shared by international financial institutions, the governments of leading industrial-financial countries, oil-producing governments, and those hardest hit of all, the non-oil-producing countries of the Third World. Among the latter, those producing other major raw materials became keenly interested in emulating OPEC—a prospect encouraged by leaders of many oil-producing countries and naturally received with mixed feelings in the capitals of industrial countries.

Of the several working groups created by the Washington Energy Conference of February 1974, the one dealing with the problem of security of future oil supplies worked on the most rapid schedule. It was only fitting that the American administration should take the initiative in resolving the problem. As the major supporter of Israel, the United States had, after all, been the intended target of the Arab "oil weapon," and yet Europe and Japan had left the petroarena with the deeper wounds. Because of its more lavish per capita use of energy, moreover, and as the world's second largest oil producer, the United States had more oil to share in time of crisis than did its industrial allies. (In 1974, the Soviet Union replaced the United States as the largest producer; third place has been held successively by Venezuela, Iran in 1970-71, and Saudi Arabia; see Table 1.) For the longer run the United States also had the most advanced nuclear technology and the largest hydrocarbon reserves (coal and shale) from which substitutes for oil might be derived.

Kissinger in his opening speech at the Washington Energy Conference left it unclear whether the United States meant the future emergency sharing scheme to extend only to oil imports or to all oil, domestic and imported. Leaving domestic oil out of account would have cut the potential American contribution by two-thirds. But that would have been quite unacceptable to Europeans and Japanese, who in effect would have been asked in advance and deliberately (as they had done willy-nilly in the winter of 1973/4) to underwrite American gambles both in Middle Eastern policy and the profligate use of energy. On the other hand, putting all available energy supplies (including coal and natural gas) into the emergency account would have more than doubled the American contribution to the pool. The statesmanlike compromise devised in the spring of 1974 was that the emergency plan would seek to equalize the supply of petroleum available to each member, whether domestic or imported—but of petroleum only.

A further crucial decision was to leave the leadership to Europeans while Americans supplied much of the staff work. The chairman of the Energy Coordinating Group that formulated these plans throughout the summer and autumn was Vicomte Etienne Davignon of Belgium, long one of the ablest advocates of political unification in Western Europe. The organization to administer the plan, known as the International Energy Agency, was established in November 1974 as a loose adjunct of OECD, with Davignon as chairman of the board, and Ulf Lantzke of West Germany as director.

The IEA treaty itself is an intricate document in the best Coal and Steel Community and Treaty of Rome traditions, with one kind of weighted majority for general policy decisions and another for decisions on emergency sharing, with one kind of scheme in effect if overall supplies are cut by 7% and another if by 10%, with careful and balanced attention to the interests of Europeans and Americans, as well as of member states which are also producers (United States, Canada, and, in the future, Britain) and of those (all the rest) which are consumers only. One of the most important features was that the scheme, once accepted by the members, would go into effect automatically—presumably by directives from the IEA to the multinational oil companies—unless any member state specifically chose to withdraw. The signal sent to the planners of any future Arab embargo was loud and clear: it would be no use trying to aim the "oil weapon" at any particular country or countries since all would stand together on the firing line so as to disperse the effect on each. (It was in the nature of the IEA's insurance plan that it would prove itself truly effective if it never came into application.) The emergency sharing scheme also looked ahead to the next item on the IEA's agenda, that of reducing consumption: any member state cutting its oil imports below 1973 levels would have that energy saving counted into its emergency reduction quota.

By the beginning of 1975, 18 countries had signed the IEA agreement, including eight members of the European Communities, the United States, Canada, Japan, Australia, New

Zealand, Sweden, Switzerland, Austria, Turkey, and Spain. France stayed aloof, because of post-Gaullist compunctions, but would be fully consulted on all decisions—in effect a sort of silent partnership in the Giscardian style. Some observers saw in the IEA a worthy successor of the Marshall Plan—an American initiative full of future potential carried to fruition in Paris and Brussels by the heirs of Spaak and Schuman; others feared a major setback to long-delayed initiatives for an energy policy for the European Communities.[29]

While one group of diplomats was busy writing insurance policies against future uses of the "oil weapon," the economic and monetary experts were devising schemes that would help finance the enormous balance-of-payments deficits expected from the price increases. The first major initiative in this field was taken by H. Johannes Witteveen, managing director of the International Monetary Fund, in January 1974, at a time when estimates of petrodollar surpluses were at their highest and fears of the deterioration of the world economy into a universal game of "beggar thy neighbor" at their most acute. Witteveen's plan called for an "oil facility" that would assist IMF members meet their "oil-induced deficits." Whereas the OPEC states were wondering whether or not to construe the consumer country negotiations leading to the IEA as an unfriendly gesture, they soon cooperated with the recycling plans of the Fund and the World Bank that, after all, would help keep their customers in the chips with which to pay for the oil. In May Saudi Arabia and Iran pledged major financial support for the Witteveen fund; on June 13 the IMF's oil facility was officially established with a planned initial capital of SDR3 billion; and by August five OPEC countries (Saudi Arabia, Iran, Venezuela, Kuwait, and Abu Dhabi) and two other oil producers (Canada and Oman) had lent it SDR2.8

billion at 7%. Before the end of 1974, loans from the oil facility had gone to Italy (27%), seven "non-industrial developed countries" (32%), and 30 developing countries (41%).

Since the problem of oil-induced payments deficits was by no means resolved at the end of the year, the IMF continued the facility for 1975 on an almost doubled scale. The conditions of the lenders have been tightened up, with loans at 7.75% and (in the IMF's circumspect jargon) "subject to stricter policy conditionality than in 1974." Specifically, any prospective borrower must submit for Fund scrutiny its "policies to achieve medium-term solutions to its balance of payments problems" and "describe any measures to conserve oil or to develop alternative sources of energy." [30]

Meanwhile the World Bank on June 13, 1975, decided to establish, over United States objections but with oil exporters' support, a new so-called "third window" to make up to $500 million in loans at 4.5% to countries unable to pay the usual 8.5% (at the "first window") or to qualify for the interest-free International Development Association loans (the "second window"). This new facility, however, was not aimed solely at difficulties arising from petroleum payments.[31]

Among the beneficiaries of these diverse international facilities of the IMF and the World Bank were those Third World countries without oil production or other lucrative exports. Indeed, for these countries—nowadays sometimes known as the "Fourth World"—international institutions were the major source of help in their dire need. Meanwhile, the major industrial consuming nations, which pay about nine-tenths of OPEC's additional oil bill, have made more specific petrodollar recycling arrangements among themselves. The first of these was a major loan made late in 1974 by Germany to Italy but put in the form of an arrangement sponsored and guaranteed by the European Communities under which any member could arrange to borrow with the guarantee of any other.[32]

In January 1975 OECD began to confer on plans to help its weaker members, and by April it reached agreement on a Financial Support Fund, of $25 billion, semiofficially known as the "safety net"—presumably for the trapeze act of petrofinance. "The principal objectives of the FSF are to encourage and assist its members to avoid unilateral moves [such as Italy and Denmark resorted to in 1974] that would restrict international trade or artificially stimulate exports as well as encourage them to follow appropriate domestic and economic policies, including adequate balance of payments and cooperative measures to promote increased production and conservation of energy."[33] There are two methods of borrowing: directly from the FSF or with its guarantee in the open market. The United States, West Germany, and Japan are underwriting a majority of the available total. Although there is no maximum limit on borrowing, access is progressively restricted by the voting procedure. A loan up to the amount of the borrower's quota requires a two-thirds majority; if the member's outstanding debt is between 100% and 200% of its quota, a 90% vote is needed (which gives any one of the three chief lenders a veto); and beyond 200% there must be unanimity.

Other regional groups have also felt the need to supplement the IMF oil facility. In May 1975 the United Nations Economic Commission for Latin America (ECLA) endorsed a proposal for a regional safety net of its own, with a $4 billion capital fund corresponding to one-third of the region's net balance-of-payments deficit for 1974.[34]

Spending and Investing

The channels into which OPEC member states themselves would cycle their funds were, of course, a crucial determinant of the size and shape of the recycling problem for everyone else. Apprehensions about the petrodollar overhang tended to mount and decline somewhat erratically, much as quotations do on any commercial futures. The oil price rises of October and December 1973 came as a major shock, and a grave view of the recycling problem was prevalent in January and February 1974. By early spring the monetary flow toward OPEC countries was found to be much less than feared—a natural consequence of the 90-day payments terms then customary between companies and OPEC governments, and of reduced consumption during the embargo. But when most payments for the first quarter of 1974 had been made, when throughout the spring oil storage tanks in importing countries were busily being filled, and when by mid-year negotiations about "participation" and "buy back prices" had raised the government take for all of 1974, retroactively, from $7.01 to $9.27, apprehension mounted steeply. Toward the end of 1974, when OPEC countries had had time to think about how to spend their new riches, had begun to translate their economic expansion plans into imports, and had generally been observed to conduct themselves in a *sotto voce*, statesmanlike fashion on the financial scene, quotations on apprehensions in the markets of opinion once again declined.

For the international banking system, the worst moments came in the spring of 1974 when the first oil company payments to OPEC under the new tax schedules were building up and while OPEC countries (like any common mortals who might suddenly have come into unexpected billions) were sorting out their ideas of what to do with the money. The result was a plethora of short-term deposits in the Eurocurrency market, at the very time when countries with the most intractable payments problems, such as Britain and Italy, were turning to the Euromarket for medium-term loans. But the first notable bank failures (Herrstatt of Cologne in June 1974 and Franklin National of New York in October 1974, the former at least due to sloppy currency dealings) served as salutary warnings. Private banking firms became careful to lower interest on short-term deposits or refuse them outright, and to increase it on longer-term ones. In London, which remains the chief center of the Eurocurrency market, the Bank of England moved, in the words of its responsible official, "in common with other supervising authorities [i.e., central banks] . . . to refine our supervisory techniques." [35]

In the second half of 1974, OPEC countries had issued massive new import orders and had had time to place their remaining funds into longer-term assets. Imports gained momentum to such an extent that early in 1975 it was not uncommon for ships at Jiddah or at Persian Gulf ports to have to wait as long as one month to unload, adding corresponding demurrage charges to the steeply rising import bill. The value of aggregate imports into OPEC countries doubled between 1970 and 1973, and tripled between 1970 and 1974. The leading importers now were Iran, Venezuela, and Algeria; and half the total imports came from the United States, West Germany, and Japan. The development plans of OPEC member states are ambitious enough to imply a growing need for imports, quite aside from the trickle-down effect of a rapidly rising standard of living. Thus the Persian Gulf OPEC members in the next five years are planning to spend more than $350 billion on economic development.[36] These sums, however, may have to be scaled down, as they already have been in Iraq and Iran, if oil revenue projections are not met or unforeseen bottlenecks develop.

Several Middle Eastern countries, notably Iran, Saudi Arabia, Iraq, and Libya, reserved substantial portions of their new oil earnings for arms purchases abroad. Middle Eastern countries (including those just listed as well as Jordan, Egypt, Syria—and foremost among all, Israel) have been among the most lavish military spenders in the world, whether military expenditures be reckoned as a proportion of GNP or as expenditure per capita. The new oil riches made possible even greater military expenditures. Between the two leading OPEC countries of the Middle East, Iran and Saudi Arabia, there has been something of an arms race for the last decade, with Iranian military expenditures increasing in direct proportion to oil income, and Saudi expenditures catching up in spurts after a year or two; the United States has been the principal outside arms supplier for both.[37]

Almost equally as newsworthy as arms sales have been foreign direct investments by Middle Eastern oil companies in major industrial concerns in the West. In July 1974 Iran purchased 25% of the stock of Krupp Hüttenwerke of Essen, Germany, for an undisclosed sum; and in November, Kuwait bought a one-seventh stake in Daimler-Benz of Ludwigsburg, also in Germany, for a reported $300-$400 million. There also were unconsummated negotiations for Iranian purchase into Ashland Oil of Kentucky, and Pan American World Airways of New York; and rather spectacular purchases by a variety of Middle Eastern buyers of real estate on the Champs Elysées, in the London financial district, of ancient British castles, and along choice South Carolina beaches. Probably the fashionable real estate market will be out of kilter for some years, or even decades, through this infusion of petromillions. But the purchases of stock in Krupp, Daimler, and others presumably are meant as blue chip investments rather than attempts to gain control of Western industry. If the Iranian-nominated board members of Krupp or the Daimler directors representing Kuwait should start showing the slightest signs of acting against the West German national interest, it would take no more than a simple cabinet decree or parliamentary law to nationalize their shares—a procedure the efficacy of which Kuwait, Iran, and others have demonstrated on less provocation. Therefore, a journalist's computation that "It would take Saudi Arabia only four months to buy up, at current depressed prices, the 30 leading companies making up the [London] *Financial Times* share index," remains no more than a bit of flamboyant reporting—not serious analysis.[38]

A sizable portion of OPEC oil earnings went into bilateral aid to Third World countries. Thus, the chairman of OECD's Development Assistance Committee has calculated that the ten aid-giving OPEC countries in 1974 spent 1.8% of their GNP on "official development assistance" (as that term is defined by the DAC), as against only .33% for OECD member countries.[39] Nearly all of this bilateral aid went from Middle Eastern countries to fellow Arabs in Egypt, Syria, and elsewhere, or fellow Muslims in Pakistan, as well as to India—whose recent explosion of a nuclear bomb, large undeveloped natural and human resources, and close proximity may constitute major points of attraction. Among the newsworthy, but financially less significant, payments were subsidies from OPEC countries to Palestinian guerrillas and to the Irish Republican Army (the latter, reportedly, from Libya).

Merchandise imports, including current consumption and items needed for economic development; arms purchases, direct investments in blue chip industry or fashionable real estate; support of international lending facilities to help the poorer customers pay their oil bills; and bilateral aid to fellow Muslim or Asian countries—these categories accounted for most of the foreign exchange expenditures of the *nouveaux riches* OPEC nations of the Middle East in 1974 and 1975. The non-import part of this total was estimated by the IMF as being $70 billion in 1974 for all OPEC countries in the aggregate. Of this, $11 billion were direct investments in the United States; $12 billion were claims on central banks in other developed countries; $21 billion were placed in the Eurocurrency market; and $6 billion given to the IMF oil facility or in bilateral aid. A further $4 billion were used for payments (on schedule or early) on past debts or advance payments for imports; and, as of March 1975, there was a further $16 billion on the books of OPEC governments as accounts receivable from oil companies for 1974—or, in other words, credit extended to the customers.[40]

What emerges most clearly from this catalog of practical responses to the financial impact of the oil price jump is that OPEC countries have recognized the need for international financial cooperation to smooth the movement of funds. OPEC countries have been careful to minimize their exposure to xenophobic reactions of the oil-importing countries by keeping their official investments in low profile. In buying for their domestic needs, the OPEC states have displayed a prudence with which many observers would not have credited them.[41] How long international financial markets will be able to deal with a growing flow of funds and whether host countries will accept increasing amounts of foreign direct investments is uncertain. Perhaps patching up the old fabric will be too difficult as quantities grow. Designs for new institutions, such as a "mutual fund for OPEC investors" with ample participation of investors from industrial countries, may have to supplement or in part replace the current arrangements.[42]

But the current (rather loose and haphazard, or "unseen-hand") arrangements have their own strength in simple economic logic. OPEC funds tend to go, in the first place, to the New York and London (including Eurocurrency) financial markets, because they are the only markets large enough to absorb the balances involved. OPEC orders for imports go primarily to the United States, West Germany, Japan, Great Britain, and France (in that sharply descending order), because those are the only countries with the industrial plants and industrial skills to handle orders on that order of magnitude. And West Germany and the United States, as major industrial powers with sizable foreign exchange reserves, also attract much of OPEC's portfolio investments. The net result is that these five countries in one way or another are much less severely hit by the oil payments gap than other OPEC customers: on balance they attract, rather than lose, petrodollars. This in turn gives these five key countries of OECD and IMF the steady flow of funds with which to preside over the rechanneling of the flow of petrodollars to the needier ones among the industrial countries. And collectively, of course, the industrial countries are the customers for 90% or so of OPEC's oil, and, therefore, 90% or so of the "recycling" problem arises among them.

With much prodding from international institutions such as the Monetary Fund and the World Bank, both OECD consumers and OPEC producers have participated, on a less than adequate scale, in easing the payments burden for those Third World countries which are least able to bear it.

Bauxite and Bananas

Even as OECD countries planned how to counteract the effects of the next oil embargo, and bank managers learned how to cycle and recycle the sudden tide of oil monies, diplomatic attention turned to raw materials other than oil. For some of the non-oil producers here was a ray of hope for coping with their particular oil payments crises through cartels of their own. For some OPEC leaders, notably the Algerians (but soon joined by Iranians, Saudis, Venezuelans, and others), here was a chance to elevate their exercise in political economy to one of moral ideology and to strengthen their bargaining position in meetings with the oil-consuming countries. And for a large majority of the United Nations here was another campaign in the long crusade of the nonaligned and the less developed countries (LDCs) against "neo-imperialism." It was in this atmosphere that the U.N. General Assembly on May 1, 1974, adopted a Declaration and Programme of Action on the Establishment of a New Economic Order, and decided to convene its Seventh Special Session in September 1975 on the subject of "Development Cooperation"; that the European Communities at Lomé concluded a convention with forty-six less developed countries of the African, Caribbean, and Pacific regions; and that the Conference on International Economic Cooperation convened in December 1975.

The decisive question with regard to other raw materials was the possibility of cartels that might emulate OPEC's success in reallocating income from the industrial countries to the LDCs that supplied their raw materials. As with petroleum, such possibilities depended not on rhetoric or intent but on objective institutional features of the world economy and on future opportunities skillfully exploited. The clamor about other cartels is an echo of the OPEC revolution, which is the proper subject of this inquiry. To examine briefly why producers of other commodities may or may not be able to follow OPEC's example will also help us understand how OPEC's own success became possible and how durable it may prove.[43]

Some physical and technical features of oil distinguish it rather sharply from other commodities. Oil is not an agricultural product that once grown must be harvested, and once harvested stored, and once stored disposed of through sale or otherwise. Oil, once drilled, either flows freely (as around the Persian Gulf) or it must be pumped (as in the United States and in Libya). In either case, its production can be regulated—within very wide limits—by adjusting the requisite number of valves or pumps. Storage has been a very minor factor in the petroleum economics of the past; and, in contrast to crops, pests or spoilage play no role at all. In the economics of petroleum there can be excess capacity but, strictly speaking, no glut. (The storage programs undertaken by members of the European Communities since the late 1960s and now under the aegis of the IEA are, of course, intended not for glut but against scarcity.) And, at the other extreme of the commodity spectrum from crops, metal ores require little precaution in storage and can be stored compactly. But oil, being liquid and flammable, must be stored in steel containers that cost as much as the oil itself—or rather, as it used to cost before the price rises of the 1970s.[44]

Oil production requires small cadres of technicians and managers. Metal mining and agriculture, by contrast, are labor-intensive; any government contemplating a shutdown in a dispute over tax or price therefore must take note of the political dissastisfactions that may ensue from sizable unemployment.

The only current substitutes for petroleum are coal and natural gas for heating, electricity, and industrial uses; and, to a far lesser extent, also hydropower and atomic energy (about 10% and 1% of total world energy consumption, respectively) for electricity. But in its transportation uses for airplanes, trucks, buses, automobiles, and nowadays ships (as well as for lubricants) and as a raw material for much of the chemical industry there are no ready substitutes for oil at all. By contrast, one metal can often substitute for another, and a (petroleum-based) plastic for either, the particular substitution depending on use and on relative prices. For such tropical fibers as hemp or jute there are many substitutes; and such tropical foodstuffs as bananas, cocoa, tea, and coffee are far from essential—which means that in effect they also have many substitutes, such as orange juice, corn flakes, or scrambled eggs. Moreover, unlike minerals which must be mined where nature happened to put them, any plant can be grown within a wide climatic zone. For all metals there is a scrap market; and the higher the price of new ore, the larger that market for scrap. In the United States and Europe these days, secondary recovery accounts for 30% to 45% of total consumption of iron, copper, and lead.[45] When a metal becomes as valuable as gold, repeated use is so common as to make the term "scrap" inappropriate—but so it is for oil: scrap petroleum becomes air pollution. Lastly, it is one of the accidents of geology that most of the world's petroleum, and all of it that can be produced for under 50 cents a barrel, has been found in Third World countries. Most metals, by contrast, are widely dispersed throughout the earth's crust; and for many of them certain industrial countries (the United States, Australia, Canada—as well as the Soviet Union) are leading producers, so that Third

World countries as a group do not dominate the market, but rather supplement American domestic production or the international trade among OECD countries. And where Third World suppliers dominate for the moment, that ascendancy might not always prove durable.

Each one of the factors listed makes the maintenance of a cartel easier in petroleum than in almost any other commodity: the ease of curtailing production, the bulk and expense of storage, the absence of substitutes in one or two sets of essential uses, the impossibility of recycling, and its geologic concentration in the Third World rather than the consumer countries. It is a formidable list that ought to give pause to any persons afraid or hopeful of a parade of cartels marching onto the stage of economic history in OPEC's van.

Other comparative advantages accrue to OPEC from being a well-established cartel and, by now, a spectacularly successful one. Cohesion in a cartel, whether based on production quotas or, as in OPEC's case, on price solidarity, is hardest to maintain at the start. As long as the cartel seems vulnerable and therefore temporary, each producer will be tempted to increase its revenue in the short run by decreasing its prices proportionately less than it hopes to expand output. This is the tendency to "chisel and cheat," which Adelman rightly identifies as the generic problem of all cartels; and of course the anticipation of the need to cut prices becomes a self-confirming prophecy, and that of gains from the operation a self-defeating one. And even during the vulnerable, initial phase OPEC had the advantage of having a majority of members from a single Middle Eastern, Islamic region.

But against these advantages of the established cartel must be set the advantages of the newcomers. Just because OPEC, after one false and one modest start, became such a spectacular success, others can benefit from its lessons without charge.

OPEC has demonstrated several important principles of cartel technique. First, price solidarity can be a substitute for production quotas, and indeed a superior one, because it will cause less contention among the cartel members—and with consumers—than would concerted production cutbacks.[46] Second, OPEC has derived great strength from its flexible negotiating tactics. Members of the cartel need not act in unison: Libya can raise the tax rate, Iran the posted price, and Saudi Arabia develop participation as a means of raising both. This allows for differences in temperament, judgment, and opportunity between one country's leadership and the next—and gives each of them by turns the chance to take the lead. But in any showdown over cash or control, it is essential that no producer undercut the other by allowing the companies to increase production, and that all of them follow the leader of the moment by invoking the principle of "the best of current practice." As Zuhayr Mikdashi has pointed out, the "relatively low level of authority that member countries provide the OPEC organization has so far not proven a source of weakness. On the contrary, it has enabled the organization to weather dramatic changes in world oil conditions and in its members' policies. It has enabled OPEC to survive periods of political division, of economic rivalry, and of mutations in international relations." [47]

Third, OPEC has shown that a network of multinational producing and trading companies can become not the enemy but the ally of a governmental cartel. As long as royalties or taxes are finely adjusted to known differences in quality and transport cost, the tax-paid cost plus transport will constitute a publicly known price floor which vertically integrated companies—without need for collusion—will enforce throughout their markets, meanwhile serving as the cartel's "tax collecting agency."

Lastly, OPEC may have demonstrated something about the likely reaction of consumer countries to cartel action: namely, acquiescence. If no military intervention and no economic counterembargoes or boycotts are undertaken even for petroleum—so it may be argued—none will be undertaken for the sake of bananas or bauxite. But here the lesson is rather ambiguous. Drastic measures, such as the October 1973 embargo and cutback, could not have been justified by the cartel members (and might not have been tolerated by the customers) without the Yom Kippur War. And countries smaller or closer to the principal Western power are more vulnerable not only to military or subversive threats but also to economic pressures. For example, a Caribbean island that depends on tourism in addition to its primary commodity might be very vulnerable to a simple measure of banning American travel there. In anticipating the reactions of importing countries, every exporting country therefore will have to assess its situation and its chances, and adapt its tactics accordingly.

There is one last, crucial difference between OPEC and other potential cartels, and that is the relative place of each commodity in the total pattern of world trade, and hence finance. In pre-oil-embargo values, roughly two-thirds of all international trade among non-Communist countries goes from one developed country to the other; about 15% each is trade from the developed to the developing countries, and vice versa; and 4% is trade among the LDCs. By commodity category, about two-thirds of total world trade is in processed or manufactured products, and only one-third in raw materials. Of the total value of raw materials in international trade, about 40% consists of food, 31% mineral fuels (including mainly oil, but also coal and gas), and only 28% of all other minerals.[48]

All this is to say that petroleum, even before OPEC's steep price increases, was far and away the single most valuable commodity traded in the world generally and from the Third World to the industrial nations in particular. In 1970 Third World exports of fuels to the industrial world ($13.95 billion f.o.b.) had nearly double the value of all metals and minerals so traded ($7.17 billion, also f.o.b.)—from copper, bauxite, and iron to sulfur, manganese, tungsten, and the rest. Thus, even if all metals and minerals exported by the Third World were joined in a tight gigantic supercartel, the financial burden on the consumer countries would be only half as great as that imposed by OPEC, and so would be any disruption inflicted on the world's economy.

The Third World producers of non-oil raw materials, in sum, will not, in the foreseeable future, approach OPEC's resounding success of 1970-75, and it would hardly be in their interests to risk any naked contest of economic power. But that very circumstance creates the possibility of achieving progress toward some kind of new international economic order through diplomacy rather than confrontation. In February 1975, the European Communities signed the Lomé Convention with 46 LDCs, most of them African and Caribbean, in which the EC yielded on its earlier demand for reciprocity of trade concessions, and also established a $450 million fund to stabilize the earnings of those LDCs from twelve primary products. The trade concessions were an indirect form of development aid; stabilization of wildly fluctuating commodity prices, while a major gain to the producers, is no loss in the long run to the consumers, either; and the funds required are small.[49]

The OPEC countries, moreover, have thrown their diplomatic support behind the demand for a series of world-wide commodity agreements. As the preliminary tripartite meetings of OECD members, OPEC members, and non-OPEC LDCs broke up in Paris on April 16, 1975, the oil countries, led by Algeria, insisted that the future agenda include the non-oil commodities, and the United States insisted that it do not. But soon there was what a State Department spokesman referred to as "a further evolution in our thinking"—stimulated in part, it was reported, by quiet pressure from Saudi Arabia and Iran. In his speech at Kansas City on May 13, Secretary Kissinger indicated that the United States was prepared to discuss "new arrangements on individual commodities on a case-by-case basis as circumstances warrant." This indeed became the basis for the Conference on International Economic Cooperation in December 1975.[50]

Kissinger's new departure ran into public criticism from the Treasury Department, which opposed both State's encroachment into trade matters and the substance of the new policy. Treasury officials therefore encroached somewhat undiplomatically on State's turf by volunteering the explanation that Kissinger's offer to consider commodities on "a case-by-case basis" was "in fact a rejection" of "broad scale commodity agreements aimed at fixing prices." Earlier, the Congress had amended the Trade Act of 1974 to deny trade preferences to any state that engages in cartel-like behavior either for economic or political reasons—which implied the denial of hemispheric preferences for Venezuela and Ecuador.[51]

These public strictures by Congress and the Treasury Department reflect a narrowly economic assessment of the practicality or effects of commodity cartels. They appear to ignore the considerable pressure that OPEC members can bring on the United States and its allies so as to extract concessions on behalf of other LDCs. The immediate question is not whether commodity agreements are desirable in the abstract, but whether the United States can muster the diplomatic support—at meetings such as those in Paris, in the United Nations, or elsewhere—to ignore the demands pressed by the oil countries on behalf of their poorer neighbors. The State Department's revised position seemed to stem from a clearer assessment of these diplomatic realities. There also is much to be said for discussing other raw materials while consumer countries can bargain from strength and make concessions broadmindedly, rather than for waiting, as in the case of oil, until the balance of economic power might tip the other way.

Meanwhile, the coffee-exporting countries have obtained Venezuela's backing to the extent of $100 million in trying to revive their earlier price-stabilizing agreement and to establish a stockpile. Such stabilizing agreements, as noted a moment ago, do not require large funds, and benefit the producers without harming the consumers. The United States and the European Communities would be well advised to consider similar positive initiatives for commodities that otherwise might become subject to "price gouging" or supply interruptions.[52]

International conferences discussing issues of importance to the industrial world have in recent years frequently been frustrated by the continual insistence of LDCs on discussing plans for a "New Economic Order." An imaginative initiative from the developed countries designed to improve the lot of some of the LDCs might vastly improve the diplomatic setting and lead to more fruitful discussions on such topics as exploitation of the seabed and control of environmental pollution.[53]

Notes

1. Letter to the *New York Times,* October 3, 1974.

2. Adelman, "Is the Oil Shortage Real?", p. 87.

3. See his address to the American Petroleum Institute, November 12, 1974, as paraphrased by *New York Times* correspondent William D. Smith (November 13): ". . . oil prices would come down, . . . it was no longer a question of whether but when." Simon reitereated this position on several occasions in the first half of 1975. His second successor as Federal Energy Administrator, Frank G. Zarb, echoed the same formulation with regard to domestic oil prices ("The Seven Truths of Energy," *Wall Street Journal,* September 10, 1975): "Truth Number Four: The Question is Not Whether Oil Prices Will be Decontrolled, But When."

4. State Department Press Release, May 27, 1975.

5. Louis Kraar, "OPEC Is Starting To Feel the Pressure," *Fortune,* May 1975, p. 186.

6. Federal Energy Administration, *Oil: Possible Levels of Future Production,* A Task Force Report for Project Independence (Washington: GPO, November 1974), pp. 2-4. Under their "Accelerated Development" case, which assumes no increase in 1974 consumption and a price of $11 per barrel of crude, the United States could be free of oil imports by 1985; yet "The limits of the U.S. resource base make it unlikely that [crude] production could be maintained at such levels for more than a few years."

7. Organisation for Economic Co-Operation and Development, *Energy Prospects to 1985,* 2 vols. (Paris: OECD, December 1974), vol. I, p. 10. The report was prepared under the direction of Professor Hans K. Schneider of the Institute for Energy Economics of the University of Cologne.

8. *Ibid.,* vol. I, p. 8 and *passim.* The higher OECD assumption was fairly close to the price of $10 to $11 actually reached at the beginning of 1975. A team of Brookings Institution specialists, also publishing their study late in 1974, blithely assumed that prices throughout the period from 1974 to 1978 would stay at the $5 to $6 range, in 1973 dollars: Joseph A. Yager, Eleanor B. Steinberg et al., *Energy and U.S. Foreign Policy* (Cambridge, Mass.: Ballinger, 1974), p. 290; this was fully one-third below the price already reached at the time of publication. Other portions of the book assume f.o.b. prices (also in 1973 dollars) of $4 to $6 for 1975-76, and of $4, $6, or $8 in 1980-85 (pp. 167, 171); by what mechanism such dramatic and early price breaks were supposed to come about is left unexplained.

9. First National City Bank, *Money International,* vol. 3, no. 5 (May 30, 1975), and *Monthly Economic Letter,* June 1975, pp. 11-15; Irving Trust Company, *The Economic View from One Wall Street,* March 20, 1975; Morgan Guaranty Trust Company, *World Financial Markets,* January 21, 1975.

10. W. J. Levy Consultants Corp., *Future OPEC Accumulation of Oil Money: A New Look at a Critical Problem,* June 1975. Table A-4 has a detailed comparison with the estimates of the three banks. Appendix B discusses the "Variations from Base Case."

11. OECD's "base case" compiles predictions of consumption, etc., made just before the 1973 embargo. These imply a price of about $2.50.

12. See *New York Times,* February 12, 1974.

13. *Ibid.,* September 24, 1974.

14. Hollis B. Chenery, "Restructuring the World Economy," *Foreign Affairs,* vol. 53, no. 2 (January 1975), pp. 255-57.

15. Alarm was given in C. Fred Bergsten's four articles, "New Era in World Commodity Markets," *Challenge,* September/October 1974, "The Threat From the Third World," *Foreign Policy,* 11 (Summer 1973), "The Threat Is Real," *ibid.,* no. 14 (Spring 1974), and "The Response to the Third World," *ibid.,* no. 17 (Winter 1974-75). Stephen Krasner, "Oil Is the Exception," *ibid.,* no. 14 (Spring 1974) and Bension Varon and Kenji Takeuchi, "Developing Countries and Non-Fuel Minerals," *Foreign Affairs,* vol. 52, no. 3 (April 1974) did not see a serious threat.

16. The interview was published in *Business Week,* January 13, 1975, but widely reported in the press by late December 1974.

17. Miles Ignotus (pseud.), "Seizing Arab Oil: The Case for U.S. Intervention: Why, How, Where," *Harper's,* March 1975, pp. 45-62; *The Economist,* May 1975, Survey, pp. 35f.

18. "Oil: The Issue of American Intervention," *Commentary,* January 1975, pp. 21-31, and "Further Reflections on Oil and Force," *ibid.,* March 1975, pp. 45-56. Earlier, a cover of *Der Spiegel* (January 13, 1975) had depicted Ford and Kissinger in U.S. Marine garb storming a palm- and derrick-studded beach.

Two other published scenarios may be mentioned in passing. Paul Erdman, writing for the irrepressible *New York* magazine ("The Oil War of 1976: How the Shah Won the World," October 2, 1974, pp. 39-51) depicted a "two-day war" whereby "the shah" (consistently set in lower case) gains "control of nearly all the world's oil supply and . . . the right to set [its] price . . . The Industrial Era around the world ended soon after." Former Senator William J. Fulbright, in a reflective essay of devastating irony ("Fulbright's

1980 Middle East Scenario," *The Washington Star,* July 13, 1975) foresaw a new Yom Kippur War of ten days' duration, an Arab oil embargo lasting well beyond the war, an American invasion of the Saudi and Kuwaiti fields in the fifth month of the embargo, a complete interruption of all Middle East oil exports for the next six months, and a further increase in the price of oil by the U.S.-led "International Petroleum Authority" to meet unforeseen costs of "security, reconstruction, and antiterrorist operations."

19. Richard N. Gardner raised this possibility in November 1973, as did eight U.S. economists (including all four American Nobel laureates). Paul Samuelson, one of the Nobel winners, later said that retaliation would not work. The U.S. government rejected this strategy because the United States does not play an important enough role in Middle Eastern imports to do it alone, and the cooperation of the major consumers would be too difficult to obtain and then sustain (*New York Times,* November 15, 1973, November 21, 1973; December 3, 1973; January 21, 1974). For a fuller exposition of Gardner's views, see his article, "The Hard Road to World Order," *Foreign Affairs,* vol. 52, no. 3 (April 1974), pp. 556-76, esp. pp. 564-68.

20. Cf. above, p. 18, and note 37 (p. 40). to Chapter 1.

21. Cf. below, p. 104 and note 22 (pp. 116f). to Chapter 3.

22. This, of course, is a non sequitur. Instead of the cartel breaking, the companies might be put in the same situation in the United States that they already play in Europe—of collecting taxes for both OPEC and consumer governments. For a discussion of the tax credit cancellation proposal see, Krueger, *The U.S. and International Oil,* p. 119ff.

23. Walter J. Levy, "An Atlantic-Japanese Energy Policy," *Foreign Policy,* no. 11 (Summer 1973), p. 190, believed that a common

consumer policy is not a "prelude to a confrontation with OPEC . . . but . . . *the only way to avoid confrontation*" (Levy's italics). Henry Kissinger, in his opening address to the 1974 Washington Energy Conference, stated that the "central purpose" of the Conference was "to move urgently to resolve the energy problems on the basis of cooperation among all nations" (*New York Times,* February 12, 1974).

24. Henri Simonet, Vice President of the Commission of the European Communities, wrote that "during the Copenhagen meeting [of EC heads of governments, December 14-15, 1973] four Arab Foreign Ministers popped up and started a round of talks with their European colleagues. This was the origin of the Euro-Arab dialogue . . ." ("Energy and the Future of Europe," *Foreign Affairs,* vol. 53, no. 3, April 1975, p. 452.) The Copenhagen declaration "confirmed the importance of entering into negotiations with oil-producing countries on comprehensive arrangements . . ."

25. The title of his article in the January 1975 issue of *Foreign Affairs.*

26. The preparatory meeting in April 1975 to plan a world energy conference was attended by four OPEC members (Saudi Arabia, Iran, Venezuela, and Algeria), three representatives of the industrialized states (United States, the European Community and Japan), and three non-oil exporting less developed countries (Brazil, India and Zaire). The December meeting included representatives of the European Community and of 26 other individual nations.

27. *New York Times,* May 14, 1975; May 15, 1975; May 28, 1975.

28. The scheme is explained in Thomas O. Enders, "OPEC and the Industrial Countries: The Next Ten Years," *Foreign Affairs,* vol. 53, no. 4 (July 1975), pp. 625-37; esp. 634f. Robert Krueger suggests that the shift of concern in the United States government

from setting a ceiling to prices to setting a floor price occurred in the autumn of 1974 (*The U.S. and International Oil,* p. 90). Although the floor price scheme was taken up by the International Energy Agency and the Commission of the European Communities, Mr. Enders' designation as ambassador to Canada in August 1975 marks the quiet end of a period of preparations for confrontation between the United States and the oil producers.

29. Perhaps the subtle parallel can be completed by noting that the Marshall Plan was announced by a Washington-based Secretary of State at a Harvard commencement, and the emergency oil scheme by a Harvard professor on leave as Secretary of State. For the first of the views mentioned in the text, see Dankwart A. Rustow, "Europe in the Age of Petroleum," in Steven J. Warnecke and Ezra N. Suleiman, eds., *Industrial Policies in Western Europe* (New York: Praeger, 1976), pp. 192-207; for the second view, the EC's commissioner for energy, Henri Simonet, who concludes his article, "Energy and the Future of Europe" (p. 463), with a gloomy quotation from Paul Valery: "L'Europe aspire à être dirigée par une Commission américaine. Toute sa politique l'y dirige." Norway, jealous as ever of her independence and even more of her new oil riches under the North Sea, had been involved in the preliminary work but in the end refused to sign; the possibility of ad hoc cooperation in a crisis was left open.

30. For fuller details, see *IMF Survey* for 1974 (March 4), p. 65; (May 6), p. 129; (June 17), p. 177; (December 19), p. 386; and 1975 (April 4), pp. 108f. An SDR (Special Drawing Right) was worth U.S. $1.21 on July 1, 1974, and $1.25 a year later.

31. *Wall Street Journal,* July 31, 1975, and *New York Times,* June 14, 1975. An additional plan, to allow countries to sell some of their official gold holdings on the open market and then use some portion of the difference between the market and official selling prices of that gold was implemented early in 1976.

32. See *New York Times,* September 1, 1974. The arrangement via the EC made the loan more palatable to opinion in both Bonn and Rome. It was the EC's first tentative venture into such a financial area.

33. *IMF Survey* April 14, 1975, pp. 97-105; the quotation is from p. 97.

34. *United Nations Monthly Chronicle,* June 1975, p. 29.

35. See George Blunden, "The Supervision of the U.K. Banking System," *Bank of England Quarterly Bulletin,* vol. 15, no. 2 (June 1975), pp. 188-94, at p. 190.

36. On trade see IMF, *International Financial Statistics,* July 1975 p. 39; OECD, *Statistics of Foreign Trade,* June 1975. On congestion in the ports and demurrage, see Levy, *Future OPEC Accumulation,* p. 7, and Enders, "OPEC and the Industrial Countries," p. 629. The delays since have increased considerably. The following sums were planned for development: Saudi Arabia $150 billion, Iraq, $75 billion, Iran $70 billion, and the principalities $50 billion, *New York Times,* July 21, 1975; cf. *ibid.,* June 17, 1975, and *Wall Street Journal,* July 3, 1975.

37. Cf. Edward M. Kennedy, "The Persian Gulf: Arms Race or Arms Control?" *Foreign Affairs,* vol. 54, no. 1 (October 1975), pp. 14-35. On the pattern of Middle East arms expenditures since the Palestine war of 1948, see D. A. Rustow, "Political Ends and Military Means in the Late Ottoman and Post-Ottoman Middle East," in V. J. Parry and Malcolm Yapp, eds., *War, Technology and Society in the Middle East* (London: Oxford University Press, 1975), pp. 386-99, and J. C. Hurewitz, *Middle Eastern Politics: The Military Dimension* (New York: Praeger, 1969), esp. pp. 106f. Statistics are regularly compiled by the International Institute of Strategic Studies, *The Military Balance* (London: ISSS, annual); the United States

Arms Control and Disarmament Agency, *World-Wide Military Expenditures* (Washington: GPO, annual); and the Stockholm International Peace Research Institute, *Yearbook of World Armaments and Disarmament* (Stockholm, annual) [title varies].

38. *New York Times,* December 8, 1974. Aside from the hazards of nationalization, the prices would of course not stay at their current depressed level if all 30 stocks were bought *in toto* in "only four months"; whether or not the free-wheeling journalist is aware of this, the Saudi financial advisers certainly are. On the Krupp and Daimler transaction, see *ibid.,* July 18 and November 13, 1974; on real estate purchases, *ibid.,* April 25, August 4, and December 8, 1974.

39. See Maurice Williams, "Aid Programs of OPEC Countries," *Foreign Affairs,* January 1976, p. 323, cf. p. 320.

40. *IMF Survey,* March 24, 1975, p. 81. Payments to governments lag three to six months behind deliveries to companies. The U.S. Treasury Department estimates that OPEC's investible surplus would be $45 billion in 1975 and that in the first six months of 1975 the United States has received $2.3 billion (9.6%) of OPEC's $24 billion investible surplus, compared to 20.4% of OPEC's surplus in 1974. The Treasury Department attributes the decline in the U.S. share to the OPEC tendency to invest in longer-term financial instruments and the low interest rates, compared to some other industrial countries, that are paid in the United States on such investments. *New York Times,* July 23, 1975.

41. But as Adelman, *WPM,* p. 207, noted long before the recent price explosion, OPEC's public behavior and "statements have been marked by sober good sense and the one big objective of increasing the [member] governments' revenue."

42. For this plan, see Khodadad Farmanfarmaian et al., "How Can the World Afford OPEC Oil?", *Foreign Affairs,* vol. 53, no. 2

(January 1975), p. 217. For a lucid analysis of the technical aspects of recycling, see Gerald A. Pollack, "The Economic Consequences of the Energy Crisis," *Foreign Affairs,* vol. 52, no. 3 (April 1974), pp. 452-71.

43. As Professor Penrose *(The Growth of Firms,* p. 140) aptly notes, "the countries producing crude oil are in a unique position; but this fact alone points out, by contrast, the unfavourable position of other raw-material producing countries . . ."

44. Adelman, writing in 1972, calculated that the capital outlay required to store a barrel of petroleum at ocean terminals such as Rotterdam is $1.25 per barrel—plus the price of purchasing and shipping the oil to the point of storage (*WMP,* p. 269). Since then the European countries, notably West Germany and France, have developed plans for storage in saltdomes, which saves the cost of constructing above-ground storage tanks, but unless the domes are close to the shore (as near Wilhelmshaven on the German North Sea) this requires added outlays for pipelines. Carroll L. Wilson, "A Plan for Energy Independence," *Foreign Affairs,* vol. 51, no. 4 (July 1973), pp. 657-75, calculates that above ground "even packed closely together a billion barrels of oil storage would take 10,000 acres or 15 square miles" (p. 672). For an earlier estimate of the costs of stockpiling oil in Western Europe, see Schurr and Homan, *Middle Eastern Oil and the Western World,* p. 81.

45. *Congressional Record,* February 18, 1975, p. E661.

46. "Fiscal and price measures, the preferred instruments of OPEC, have certain advantages over pro-rating and output restrictions. They are more acceptable to the rest of the world . . . . consumers generally prefer the freedom of buying any desired quantity at given prices than restrictions on the quantities offered." Robert Mabro, "Can OPEC Hold The Line?", *Middle East Economic Survey,* vol. 18, no. 19 (February 28, 1975.

47. "The OPEC Process," in Vernon, ed., *The Oil Crisis (Daedalus)*, pp. 203-15, at 208. Among those who have stressed the importance of OPEC's cooperation with a vertically integrated industry are Penrose *(The Growth of Firms,* pp. 209, 239) and Raymond F. Mikesell, *Nonfuel Minerals: U.S. Investment Policies Abroad,* The Washington Papers, III, 23 (Beverly Hills: Sage, 1975), pp. 28f.

48. Trade figures in this paragraph and the next are calculated or taken from *United Nations Statistical Yearbook 1971* and *1973,* passim.

49. For an analysis of the prospects of non-oil cartels that coincides largely with our own, see Varon and Takeuchi, "Developing Countries and Non-Fuel Minerals"; see especially p. 509: "OPEC countries . . . were dealing from a unique position of strength—in that they had no major conflicting trade interests, either domestic or within the group, enjoyed a high degree of independence from developed countries, and came to hold large financial reserves." But they conclude that for bauxite the prospects for a "producers' alliance" are exceptionally good. Similar views are expressed by Mikesell, *Nonfuel Minerals,* p. 31, and Philip Connelly and Robert Perlman, *The Politics of Scarcity* (London: Oxford for RIIA, 1975), passim, esp. p. 140.

50. For Kissinger's speech at Kansas City, see *New York Times,* May 14, 1975; cf. *ibid.*, August 19, 1975.

51. On Venezuela, cf. Norman Gall, "The Challenge of Venezuelan Oil," *Foreign Policy,* no. 18 (Spring 1975), pp. 44-67.

52. The President's Council on International Economic Policy examined 19 "major industrial raw materials" and found a "significant vulnerability" to either "price gouging" or interruption of supply for chromium, platinum, and bauxite. The Commission of the

European Communities made a similar study and is most concerned about the supply and price of copper, lead, and zinc. For the complete analyses, see *International Economic Report of the President,* March 1975; and Commission of the European Communities, *Communication from the Commission to the Council on Raw Materials in Relations with the Developing Countries which Export Raw Materials,* May 21, 1975, Annex 1.

53. A first step in this direction was Secretary Kissinger's speech, delivered by Ambassador Moynihan, to the United Nations General Assembly in September 1975.

[THREE]

Bold but Limited Purposes

POLITICAL AND MILITARY HAZARDS

Whatever OPEC's future as a spearhead for other cartels, its own prospects will depend on the structure of the international petroleum market.

OPEC's own momentum—and inertia on the part of its customers—now are on OPEC's side. The oil producers won spectacular victories in the early 1970s by boldly exploiting their bargaining position; their very successes have strengthened that position. Predictions of OPEC's demise from political causes tend to gloss over the economic realities. It is idle, for example, to expect that internal political differences will break the cartel. OPEC is a limited-purpose alliance of governments bent on maximizing their medium-term monetary returns from oil. Deriving strength from that limitation, it has survived the most startling political anomalies. Iran continued to send oil to Israel through three Arab wars; for years it supported the Kurdish rebels in Iraq. Territorial claims have been pressed by Iraq on Kuwait and by Iran on Bahrain. No love is lost between monarchs of the Arabian peninsula and ideologues at Algiers, Tripoli, and Baghdad, or indeed between one Arab "revolutionary" ideologue and another.[1] Iranians and Saudis, as mentioned before, for some years have indulged in a multibillion dollar arms race across the Persian (or, as the Saudis would insist, Arabian) Gulf. Venezuela and its Middle Eastern partners are, geographically and culturally, poles apart. *Eppur si muove*–"And yet it moves." [2] Political divergences notwithstanding, OPEC countries since 1970 have made about a quarter of a trillion dollars off the rest of the world. Their limited partnership has paid handsomely indeed. At OPEC's first meeting of heads of state, in Algiers in 1975, Houari Boumédienne as the host helped bring the Shah and the representative of Iraq to a compromise settlement of their countries' long-standing feuds about Kurdistan and the river frontier along the Shatt al-Arab.[3]

Diplomatic relations of OPEC with oil-importing countries bear as little on the future of the world's oil as do political relations among its members. The pro-Arab declaration of the Common Market ministers at Copenhagen in December 1973 did not save Europe from a curtailment of energy supplies more severe than that in the United States. France's pro-Arab diplomacy, ostentatiously cultivated since 1967, did not forestall nationalization of French oil interests in Algeria or Iraq, earn a rebate on oil imports,[4] or secure favored status for French exports. Nor did King Faisal's and the Shah's anti-Communist sentiments and habits of friendly diplomacy with the United States save Americans from embargo or price rise. The same will be true of the "Euro-Arab dialogue" which has been laboriously prepared since early 1974. As producers and consumers of much of the world's petroleum, as importers and suppliers of large amounts of industrial equipment, and as depositors and managers of sizable funds, Arabs and Europeans will have much to talk about in years to come. But such dialogue is best carried out in many different contacts well below the ministerial summit, and any special dispensation with regard to oil is an improbable topic for the agenda.

If OPEC will not be charmed out of its billions, neither is it likely to be bullied. An embargo on food exports, to be plausible at all, would have to be joined by not only the United States but also Canada, Australia, and the European Communities—a solidarity without recent precedent. And if it is recalled that OPEC countries import only a small proportion of their food,[5] but that Europe and Japan rely on OPEC for 60% to 75% of their energy, and that the Soviets, despite recurrent agricultural troubles, might always manage some dramatic emergency shipments—then it is easy to anticipate the outcome of such an oil vs. food confrontation.

Military occupation of some of the major oilfields would be more serious. Of the two sites proposed by the magazines, the al-Hasa district of Saudi Arabia would be more manageable than Libya, because the oil fields lie on the shore rather than hundreds of miles inland, contain six times the reserves, and flow without need for pumping. The landing beaches offer few difficulties comparable to Gallipoli in 1915, Normandy in 1944, or South Korea in 1950. The defending forces are negligible, and the desert terrain holds none of the hazards of a Southeast Asian jungle. Even if saboteurs did their worst, production of oil could resume in months, and its sale at $5 or $7 a barrel might sway much world opinion after the event.[6]

But the crucial diplomatic context is that before and during a military expedition. It is hard to conceive of a quid pro quo that would make the Soviets promise neutrality, and harder to imagine them keeping such a promise. Other OPEC countries would almost certainly go on a sympathy shutdown,[7] so that, until the Saudi wells are repaired, the invading power would require sufficient stocks of oil for its military needs and for the minimal civilian needs of itself and its friends throughout the world. Military occupation, of course, would have to last for a long time—so long, in fact, as to constitute a colonial regime. What would be the sense of an elaborate exercise that after a year or two merely landed everyone on square one?

If the costs of an oil invasion will not give pause to the scenario players, perhaps its benefits will. The United States generally is assumed to be the power that could or should contemplate the expedition. But Europe and Japan consume nearly two-thirds of OPEC's oil, the United States just over one-sixth, and each pays its proportionate share. Why in the name of *realpolitik* should Americans face global nuclear war and reversion to colonialism just to improve the economic position of their Japanese and European competitors? The only situation in which an invasion, or threat of invasion, would make sense is one where the oil countries already had done their worst by way of embargo and cutbacks—that is, the "actual strangulation" scenario of Kissinger's *Business Week* interview.

Short of possible military action, the most urgent precaution is that of an emergency energy-sharing plan among the importers. This is why the agreement worked out by the International Energy Agency in the fall of 1974 remains the wisest and most appropriate element in the Nixon-Ford-Kissinger administration's vacillating energy policy.

Drastic military or political measures such as boycott or invasion have not unnaturally suggested themselves because OPEC's momentous gains resulted from the Yom Kippur War and the Arabs' use of their "oil weapon." Indignation at Israeli occupation of Arab territories and Saudi desire to help

Egypt justified the embargo and the cutbacks; and these made possible the price jumps of 1973. But if the Yom Kippur War supplied the drive and the occasion, it was the previous constellation of the petroleum market that opened the opportunity. Earlier Arab-Israeli wars had entailed belated and ineffectual embargoes—belated because the Israelis controlled the timing and ineffectual because the Arabs did not have the financial reserves that they were to accumulate by 1970. (Perhaps it was such earlier failures—or else the slosh of Watergate—that made the American administration disregard the warnings of a Saudi embargo in case of renewed Arab-Israeli conflict.) [8] By 1973 the Arabs had ample foreign exchange, their plans were coordinated, and they controlled the timing. But note that it was the Shah who insisted on the larger of the two price increases over Yamani's initial objection, and that both increases were joined by all OPEC countries. The political motivation transcended Arab hostility toward Israel and its American supporters. Conversely, as Professor Adelman has noted, "If the Arab-Israeli dispute were settled tomorrow, the producing nations would not slow down for a minute their drive for ever-higher prices and taxes." [9] The temporary factors of the Yom Kippur War came into play against a background of economic opportunity and a wider and deeper sense of grievance.

Or, to quote one of America's most perceptive oil economists, John H. Lichtblau, "The Arab-Israeli dispute has aided the psychological underpinning of this staggering development. But essentially it was a classical revolution designed to seize control from the established powers and to effect a radical redistribution of wealth. The principal reasons for OPEC's enormous success were its good organization, its high degree of cohesiveness . . . and, above all, the fact that the oil importing countries were unable or unwilling to resist OPEC's demands. Each time the producers' organization tested how far it could go, it found that it could more or less set its own limit. Only now are there signs that this situation is beginning to change—perhaps." [10]

OPEC's leaders were born at a time when their countries were colonies, mandates, or protectorates of Western powers. They have seen the colonialists depart and the multinationals arrive. They have seen Third World countries invaded: Suez in 1956, Cuba in 1961, the Dominican Republic in 1965, not to mention Vietnam. They have seen some of their own or other Third World governments overthrown with the alleged connivance of the CIA. The sense of injury throughout the Third World is massive and profound, and it finds nowhere so clearcut an expression as in the struggle for cash and control in the economy of raw materials. As a Kuwaiti minister told an American correspondent during the embargo: "We were exploited by the industrialized countries for decades. They took our oil at a very cheap price and sold us manufactured products at a very high price. It isn't only oil but most of the raw materials they took from the developing countries." So, if oil prices hurt, "let them come and talk reason with us." [11]

A balanced assessment of the interplay of political and economic factors, given by Campbell and Caruso a year before the embargo, remains equally valid today: ". . . the real challenge to Western oil interests by the producing countries is not a result of political emotion or spite at a time of crisis . . . it represents a well-calculated economic policy which fits into a long trend . . . The motivation is nationalist, anti-imperialist, anti-Western if you like—political in that sense—but the real aim is to exploit bargaining power to gain a greater share in the returns from enormously rich resources. It is in the world-wide pattern of what eventually happens to extractive industries owned by foreign companies in less developed countries." [12]

"Bargaining power" is the key. Whether the price of oil will rise, fall, or remain steady depends primarily on economic realities and only secondarily on political combinations. World demand for OPEC's oil in mid-1975 was back to 1972 levels, about one-fifth above 1970. Five years ago, Libya, Saudi Arabia, and others had enough foreign exchange to face prolonged shutdown without hardship; today those exchange reserves have grown tenfold. Emergency stocks in the consuming countries, by contrast, still correspond to no more than two or three months' consumption. While the basic pattern of demand for OPEC's oil lasts, any full confrontation between producers and consumers—an embargo vs. a boycott—could only lead to the consumers' defeat. The oil countries probably will never again bring about a drastic jump in revenues as in the autumn of 1973; they hardly would have done so at the time if the attention of the consumer governments had not been diverted by an Arab-Israeli war and the danger of nuclear showdown among the superpowers. But there is no obvious reason why OPEC should not maintain its price structure and, in favorable circumstances, obtain the unpublicized, gradual increases that in the calendar year of 1974 raised tax-paid costs by 40%.

OPEC derives added strength from its recent partnership with the major oil companies. The negotiations at Tripoli and Tehran in 1970-71 were the last instance of major confrontation between the two. The result was a sweeping victory for the governments on the issues of cash and control, and the companies, as veteran commercial realists, quickly adjusted to the situation. The stakes at Tehran were tax increases that ultimately would have cost $3 billion or $4 billion per year.

Since then, financial issues of many times that magnitude (e.g., that of 25%, 60%, or 100% participation) have been left dangling for months at a time—in the certainty that, when companies and governments agree, consumers will pay. Meanwhile the seven majors continue as OPEC's chief production managers, sales agents, tax collectors, and adjusters of production quotas—and their profits have substantially risen under the new arrangements.

Chisel and Cheat?

Predictions of OPEC's downfall from economic causes rely on a decline in demand, increase in supply, change in government-company relations, or development of alternative fuels.

There is no evidence so far of any decline in demand that would threaten the cartel's survival. The price jump of 1973-74 resulted in the first drop in OPEC exports. In the spring of 1974, as stocks were being rebuilt after the embargo, OPEC's exports rose, but not to the record level of September 1973. Then, as the world-wide recession slowed economic activity and consumers began to respond to higher prices, demand continued to shrink, going as low as 25.8 mb/d in April 1975.

For 1980, the forecasts predict OPEC exports ranging from this 25 mb/d level upward to 32 mb/d.[13] These figures assume that Alaskan production will no more than make up for the decline of oil output in the lower 48 states, that Great Britain will become self-sufficient on its half of North Sea oil, and that much of the Norwegian half will be exported. Most experts anticipate a slow but steady increase in consumption, at about 3% a year—sharply reduced from the 7% to 10% of the decade 1963-73. This means that new sources of oil will keep pace with increase in demand, slowing down dependence on imports rather than reversing it.

Forecasts for 1985 cover an even wider range of 24.5 mb/d to 36 mb/d.[14] Once again, new discoveries (mainly in various offshore areas) and the development of alternative sources, such as nuclear power and various uses of coal, are expected to absorb increases in demand rather than reduce OPEC's total market. New oil production outside the present OPEC area does not imply a lower price: Third World producers, as we saw, have always joined OPEC, and other sources, such as Canada, the Soviet Union, and United States domestic producers, have followed this price lead. The more likely effect of new production would be not a price cut but merely a division of demand among more exporters.[15]

The 25 mb/d that are the lower limits of these forecasts for 1980 and 1985 correspond neatly to OPEC's mid-1975 exports. The decline from 1974 export levels still implied a slight diminution of revenues until the price increases of September 1975. Even if these should result in a further reduction of demand, OPEC still can look forward to revenues 6 or 7 times as large as the same volume yielded in 1972—hardly a prospect for acute penury. Furthermore, by just going back to each country's lowest monthly production since the fall of 1973, total output could be cut by another 4 mb/d to 21 mb/d. It is hard to see how the industrial countries can hurt OPEC by reducing consumption without hurting themselves more.[16]

But in fairness to Kissinger, Adelman, and others, let us suppose for the next page or two that there will be consumption cuts, new oil finds, development of alternatives, and "ambitious spending programs" that together bear down hard on "the cartel's weak spot" of excess capacity. What would the actual scenario of OPEC members "under pressure to increase sales" (Kissinger) resorting to "chiselling and cheating" (Adelman) look like?

OPEC, as noted, controls its market by setting prices rather than production quotas—a smoother and less invidious procedure. The core of the arrangement has been price cohesion among Venezuela and the four major Persian Gulf producers. The other eight OPEC members, commanding no more than 28% of production, have often charged higher prices because of better quality or proximity to markets; at times they have charged even more than these objective criteria would warrant. Before long, falling tanker rates or reduced offtake by the producing companies have forced such excessive prices more nearly into line. The ever-hopeful American press rarely fails to trumpet such price reductions as so many auguries of OPEC's imminent collapse.[17] But remember that the prices of these peripheral producing countries, even after downward adjustments, are still above the standard price charged for 72% of OPEC's output, and that, considering the quantities, a $1 per barrel cut in Libya or Indonesia means an average drop of 4¢ for all of OPEC, and a corresponding cut in Ecuador a drop of less than a penny.

Although OPEC itself has never allocated production quotas, some member states have set their individual limits—typically not for reasons of cartel policy but to prevent geologically premature depletion or to preserve a vanishing economic asset for future generations. The quota-setting group in OPEC includes Venezuela and five Arab countries with sparse populations. The group that has not set quotas, and some of whom are eager to expand capacity, includes (not surprisingly) the countries with the largest populations, either absolutely as Indonesia and Nigeria, or in relation to oil income as Algeria and Ecuador. Per capita oil revenue in the first group averages over $2,000, in the second group around

$150. The split-the-cartel tactic is designed to accentuate this division between OPEC's Haves and Havenots.

The scenario of OPEC's impending collapse sees the Havenots responding to weakening demand by lowering prices (through hidden discounts, easy credit terms, and the like), thereby increasing their production and their short-term income. The Haves, after condoning as much of this cheating and chiselling and loss of their markets as they can stomach, respond in kind. The downward spiral is on, the importers' nightmare of sky-high prices is over, and the consumers can breathe a sigh of relief.

Or can they? Before they do, they had better glance at the excess capacity figures. The Havenots in April 1975 could throw an extra 2.8 mb/d on the market. But the Haves also dispose of excess capacity. By raising production just to current legal limits, they could throw an extra 3.5 mb/d on the market; by opening their valves fully, they could raise production by as much as 9.6 mb/d. Of course we do not know precisely how much of a price cut would be required to absorb these extra quantities so as to clear the market of OPEC's full capacity of 38.1 mb/d. Pre-1973 estimates foresaw such a level of consumption in 1979 or 1980 if the price had remained steady.[18] But, of course, the world has become wary of OPEC and would respond more cautiously now. We have no exact data on the price elasticity of the demand for OPEC oil; all we know is that some kind of inverse correlation obtains: the lower the price, the higher the demand. Let us therefore leave the price factor variable and also grant the assumption that, within limits of productive capacity, OPEC countries are free to sell as much oil as they please. What would happen to the incomes of Haves and Havenots as production went to the limit and the price fell by 10%, 20% . . . or 50%? (See Table on next page.)

Revenue of OPEC "Haves" and "Havenots"
at Capacity Production and Various Price Levels*

	Production (mb/d)		Annual Revenue ($ billion)					
	Current	At Capacity	Current	With Production at Capacity and Price Dropping by:				
			(1974)	10%	20%	30%	40%	50%
"Havenots" (Algeria Ecuador, Indonesia, Iran, Iraq, Nigeria)	11.7	14.5	49.3	54.8	48.7	42.6	36.5	30.5
"Haves" (Abu Dhabi, Kuwait, Libya, Qatar, Saudi Arabia, Venezuela)	13.5	23.2	47.8	73.9	65.7	57.5	49.2	41.0

* Figures for current production (April 1975) and estimated capacity (mid-1974) are from *Petroleum Intelligence Weekly*, May 29, 1975; for 1974 revenues, from *The Petroleum Economist*, March 1975 ($7.5 billion accrued to Saudi Arabia in 1974 but not paid until 1975 has been added to the figure there).

The price cuts, in other words, would hit the Havenots and the Haves differently. If the price dropped by 10%, the Havenots would come out ahead; but if it dropped by 20%, their capacity production would earn them less than their restricted production does at current prices. The Haves would still come out ahead as the price went down by 40%. But remember that it is the Havenots in their quest for larger markets and additional revenue that were supposed to start the internecine downward spiral that kills OPEC.

Realistically, of course, all producers would not (as the above table implies) expand to the limit at once. Rather, the Havenots would cut prices and increase production gradually, and the Haves retaliate to the extent required. But the basic correlation still holds: the Haves can outmatch the Havenots during every round of the game; and, once the price drops by more than 10%, the Havenots would be reducing their own income and enhancing that of the Haves. One major factor that keeps the cartel in balance, therefore, is the vast excess oil-producing capacity of the countries, foremost Saudi Arabia, that also have the greatest surplus of funds. This long-run stabilization of the cartel is a direct benefit of Aramco's and Saudi Arabia's vast program of expanding capacity from 6.5 mb/d in late 1972 to 10.5 mb/d in 1975.[19] As Sheikh Yamani has explained rather bluntly, "To ruin the other countries of the OPEC, all we have to do is produce to our full capacity; to ruin the consumer countries, we only have to reduce our production." [20]

If the effect of expansion to capacity and of price drops is calculated for each country, it turns out that Venezuela and Qatar still come out ahead at a 20% price drop, Saudi Arabia and Kuwait at 40%, Abu Dhabi at 50%, and Libya at 60%. Among the Havenots, Iran, Iraq and Algeria could not risk going full tap if the price dropped by more than 10%;

Indonesia could see the price down by 20%, and Nigeria and Ecuador by 30%.

But, as we noted, the actual behavior of the countries that theoretically can afford to play longest at the game of antimonopoly (cutting prices to raise production) has been the very opposite of what the "break-the-cartel" scenario would suggest. Abu Dhabi and Libya among the Haves and Indonesia and Ecuador among the Havenots have intermittently been charging the highest rather than the lowest OPEC prices in 1974-75. When feeling a thirst for higher revenues, they have not reduced prices, undercutting their OPEC partners for an illusory egotistic gain. Rather they have raised prices, presumably in hopes of having their partners adjust other tax rates upward. These hopes were justified by OPEC's decision of September 27, 1975, to raise the "market" price by 15%—a move followed by all members except some which had already been charging more than corresponded to their transport and quality differentials.

Secretary Kissinger's vivid scenario of "individual producers . . . under pressure to increase sales or . . . refuse further production cuts" precipitating a downward spiral thus remains unconvincing. Two other things are notable about that scenario.

First, the hopes for internecine strife within OPEC were inspired by fears (expressed just as vividly in Kissinger's speech of February 1974) that OPEC's astronomic accumulations would plunge the rest of the world into economic disaster. But there is an implicit contradiction between the earlier fear and that later hope. World-wide recession, it was feared, would result from OPEC's inability to spend its funds; internecine price cutting, it later was hoped, will result from OPEC countries overspending those vast funds and thirsting for more. Reality meanwhile steered a middle course. OPEC countries have made enough money off the industrial world to inflict some real damage; but they have not overspent to the extent of being tempted to cut their own prices and, like Samson, bring their own temple crashing down on their heads.

Second, Kissinger's phrasing presupposes (along with the reporting in most of the American press) that it is "individual producers"—i.e., the OPEC governments—that determine month-to-month levels of production. In fact, as we have seen, Venezuela, Kuwait, and Qatar are the only governments to do so—aside from temporary political cutbacks as in the 1973 embargo. At most times and places, it has been the companies, foremost the seven majors responding to changing demand in world markets, that have in fact set each country's production level from month to month.

Professor Adelman stresses this nexus between companies and governments. "Supply and demand," he has repeatedly emphasized, "have nothing to do with the world price of oil; only the strength of the cartel matters"—and the cartel's cohesion is insured through the offices of its loyal servants, the companies. (According to Adelman, this has been the reason that Saudi strategist Sheikh Yamani is opposed to OPEC governments displacing the oil companies in their downstream activities.) [21] To break the link between OPEC and the seven majors, he now proposes a plan whereby a U.S. federal agency would allocate oil imports by sealed bids.[22] Such a procedure would make possible (though not guarantee) the elimination of the oil companies from the world-wide distribution of oil. Adelman, at any rate, hopes that OPEC countries or their anonymous agents, lured by the vast American market, will submit truly competitive bids that will break the present high prices. He overlooks the possibility that all OPEC members will submit identical bids—or, worse, no bids at all. Why would the principals of the world's most successful cartel shrink from elementary collusion? [23]

The break-the-cartel scenario, in its Kissinger, Enders, or Adelman variants, has OPEC countries pressed for money and charging less for their oil. But at least two other avenues are open. The Haves can lend money to the Havenots to keep them from cheating. (Adelman cites such an instance: Arab neighbors offered a loan to Iraq in 1972 when it was tempted to sell nationalized oil at cut-rate. Another is the 1975 offer of a loan by Venezuela to Ecuador, whose high prices had resulted in reduced company offtake.) [24] By this device, the pressure among OPEC countries is equalized, and the gap between Haves and Havenots narrowed.

Or else producing countries can try to make more money by the rather obvious device of *raising* prices. The limit to such a maneuver, for the time being, is elasticity: consumer resistance to skyrocketing prices. Just what this elasticity factor is, it is hard to tell. The general level of economic activity in its cycle from prosperity to depression, the excise taxes imposed by governments to reduce consumption, the mildness or severity of the coming winter, new finds of petroleum, development of alternatives—all these impinge on the demand for oil so heavily that it is impossible to isolate the elasticity factor. (In real life economics, the *cetera* never are *paria.*) But we noted that OPEC today is selling, at six or seven times the price, the very same amount of oil it sold in 1972—which would suggest that elasticity is well below 1.0, and that additional monopoly profits stand to be made by raising the price even further.

In pushing their monopoly profits upward in the past, OPEC leaders have often observed that excise taxes on gasoline and other petroleum products in European countries roughly have equalled the income taxes and royalties levied by OPEC on crude oil at the source.[25] These European taxes, as well as President Ford's short-lived $2 levy of mid-1975, have helped OPEC to test the price elasticity of the demand for oil. If the consumer governments (so OPEC spokesmen have implied) really fear that higher prices for oil products will cause recession or other major dislocations, why not cut their own excise taxes? This suggests a possible OPEC strategy of transferring further revenues not from the consumers themselves but from their governments. If pursued by OPEC (and acquiesced in by the importing governments), this would make possible a rise in the basic price of oil from the current $11.50 a barrel to $13 or $22 a barrel, depending whether the taxes in the United States or in Europe are taken as a guide.

Economists and diplomats speculating on OPEC's downfall or eager to hasten it by policy measures have pinned their hopes for the medium term on decreased demand in the industrial countries, increased supplies from outside OPEC, or OPEC's own needs for additional funds. None of these, as we have seen, are likely to be of the magnitude or even in the direction that would warrant such optimism. The basic facts are that the world's economy in the last generation has grown addicted to oil as its chief fuel, and that the geologists have nowhere found a pool of it as large or accessible as around the Persian Gulf. OPEC's ascendancy rests on firm control at the Gulf. Unless a comparable pool is discovered somewhere else, preferably in the jurisdiction of the consumer countries, or oil itself is displaced as the world's prime fuel, that ascendancy is likely to last.

The Price of Substitutes

The gradual displacement of oil by various cheaper substitutes, of course, has been the long-term hope of those same economists and diplomats. Since the 1973-74 crisis, steps have been taken to speed that process of substitution. Great Britain, which still has the largest coal industry in Europe, has decided not to shut it down but to open new mines; so has West Germany, where the shutdown had been nearly complete. Expansion of atomic power is proceeding rapidly in France and other European countries, and more slowly in the United States. Research and experimentation are underway into such new energy sources as coal gasification, coal liquefaction, oil retorting from shale, as well as into solar, geothermal, and tidal power. With so much ingenuity applied in so many directions, major breakthroughs may well be expected in coming years.

This still leaves the major problems of timing and cost. Only coal and uranium are so technically advanced as to provide ready substitutes for most uses of petroleum, although nuclear power since World War II has been subject again and again to frustrating delays. On the assumption that the environmental problems can be solved to the satisfaction of those concerned—and that remains a large group and a large assumption—it will still require a decade until coal and uranium will furnish substantial alternatives to oil, and another decade or more until they can begin to replace it. As we saw, even the optimistic OECD estimates see coal, uranium, and oil from Alaska and the North Sea mostly absorbing additional energy demand through 1985, rather than reducing the demand for oil imports.

Whatever the uncertainties of timing, cost, in the end, will be decisive. It was asserted at the beginning of this essay that the price of substitutes sets an upper limit to any potential monopoly gains: no consumer will buy a dearer fuel if he can get a cheaper one. But there is room for doubt about how effective this limitation will prove in the case of petroleum in the future. The proposition that the cheaper substitute will displace the more expensive one belongs, after all, to that theory of the pure market under which no cartel should have formed to begin with. In an oligopolistic market, while no consumer will pay for dearer instead of cheaper energy, it also is true that no producer will sell his energy at a lower price if he can get a higher one. And indeed our survey, early in this essay, of the environment into which OPEC made its entrance well illustrates how the more expensive alternatives often set the price.

OPEC, after all, was not the first to set world oil prices above production cost; it only widened an existing gap. In the last three decades enough oil was available at the Persian Gulf at a production cost of 10 cents to 15 cents a barrel to satisfy any foreseeable world demand. In a perfect market, this oil would rapidly have driven out domestic oil in the United States (produced at $1.50 to $3 a barrel) and domestic coal in Europe (produced at the oil equivalent of $4 a barrel). But various countervailing forces intervened. Companies with world-wide oil interests did not like to see their American production undercut by Middle Eastern oil. Governments were under political pressure, from domestic oil producers in the United States and from powerful miners' unions in Europe, and also were wary of dependence on imported energy. The governments thus had strong incentives to protect the prices of the domestic fuels, and the companies were not averse to setting the world oil price at or near those levels. The companies' solution was the Galveston basing point. The governments' answer was oil import quotas in the United States (after Galveston lost its status) and a variety of taxes and subsidies in Europe that raised the prices of both coal and of oil products. Persian Gulf oil, as a result, still acquired a market, up from 1.5 mb/d in 1950 to 13 mb/d in 1970, but at a price twenty times above the production cost. It was the dearer rather than the cheaper alternative that exerted the greater influence on the world price for energy.

Meanwhile the governments of oil-producing countries were pressing for larger shares of the economic rent implicit in these prices, first at the expense of the Western treasuries (under the fifty-fifty tax credit arrangement), then of the companies, and at last of the consumers. And they pressed so successfully that the price of oil came to be not 20 but 100 times the cost of production.

Today, therefore, the roles are almost precisely reversed. Oil from the Persian Gulf and other OPEC countries is no longer the cheapest but (once the payments to the governments are included) the dearest of the fuels on the world market. But this implies the same danger as in the 1950s and 1960s, that the price of energy will be influenced more by the dearest than by the cheapest fuel. In the late 1970s and 1980s, this means that the cost of coal, nuclear energy, non-OPEC oil, and other sources, instead of setting limits to OPEC, may all rise close to the OPEC level.

Indications of this possibility are not hard to find. As noted earlier, Canada, the Soviet Union, and other non-OPEC oil producers have all followed the OPEC lead. American domestic production has been under a complex regime of partial price controls since 1973, under which the average price of domestic oil has gone from $3.39 in 1972 to $8.38 by mid-1975.[26] If controls were lifted altogether, the expectation is that it would go to $13, the very level of the cost of imports.

Just before the 1973 embargo, Professor Carroll Wilson of MIT, and formerly general manager of the Atomic Energy Commission, advocated his plan for energy independence for the United States, whereby (a) the price of energy would be allowed to rise to $6 a barrel of oil or its equivalent, (b) stringent energy conservation measures would be imposed, (c) a ten-year program of publicly supported investment on the order of $75 billion or more would open up new energy resources, especially in coal gasification, and (d) the need for oil imports for 1985 would be kept down to about 3 mb/d.[27] In short, according to informed pre-embargo opinion, the United States could purchase its energy independence within about a decade by barely doubling the price of petroleum.

Three months after publication of Wilson's article, the price of domestic oil, following the embargo, started its climb far beyond the $6 level he advocated. Yet in this same period, United States domestic production in 1973-74 declined from 11 mb/d to 10.5 mb/d, and the estimates of proven reserves recoverable under current technological and economic conditions, as compiled by the American Petroleum Institute, continued their steady decline from 43.1 billion barrels (1972) to 41.8 billion (1973) and 40.6 billion (1974).[28] A doubling or near tripling of the price had not had the effect expected by policy makers of increasing production or of augmenting future supplies. More likely, the prospect of still higher prices had the all too human effect of delaying production and of reducing the estimate of available resources.

With regard to production of alternative fuels, the American scene since the fall of 1973 has produced much talk and little action. One contribution to the delay has been the Nixon and Ford administrations' predilection for developing alternatives only through private enterprise, and the legitimate doubts of corporate planners as to what the ultimate price of energy on the world market would turn out to be. As President Ford himself explained to a group of *New York Times* reporters, "Now, most people know, who are openminded about it, that you aren't going to . . . stimulate additional drilling and production in old oil if it is $5.25 a barrel. They just aren't going to do it. You wouldn't, I wouldn't, so you have to give them an opportunity to get a better price if you are going to get them to invest their money and get more old oil." [29] The same argument, a fortiori, applied to the development of alternative forms of energy, where the required investments may be expected to be far greater than in the expansion of existing oil production. Thus, Thomas Enders, writing only two years after Professor Wilson, estimated the required total investment toward energy independence as being "in the trillion dollar range (in 1974 dollars)," over the same decade envisaged by Wilson.[30] Ironically, Secretary Kissinger's concern in February 1974 had been to get the price of OPEC down from its estimated level of $7 so as to forestall worldwide economic collapse; but only a year later his concern became to set a floor price to oil (presumably at this very same $7 level) so as to safeguard the profitability of investments in expensive alternatives.

But, as Enders himself notes, such elaborate guarantees of profitability (or against what, in current Washington jargon, he calls "downside risk") are required only where new energy sources are developed by private enterprise. By contrast, "To the extent that the financing of new energy sources is provided through government channels, such capital investment need not be seriously affected by changes in the overall price of oil, or of energy in general." [31] It is rather surprising, therefore, that he wishes to restrict government investment to " 'nonconventional' energy sources such as solar and geothermal power, oil shale, or the Canadian tar sands" rather than extend it to the equally nonconventional procedures of coal gasification and liquefaction or shale retorting, or continue it for uranium.

In the 1930s the United States government undertook a sizable program of public development of new energy sources through the Tennessee Valley Authority. In view of our far more urgent national need for new sources of energy today, it may be time to consider creating a present-day equivalent of the TVA for the production of synthetic gas or oil from coal. The question is not just one of this or that administration's predilection for public or private enterprise. Rather, it is a matter of insuring that investments in alternatives to petroleum are made at all and are made rapidly. If the key factor becomes (in the President's words) industry's "opportunity to get a better price if you are going to get them to invest their money," there is danger that that "better price" will rise to higher and higher levels—and so, in reciprocal response, will OPEC's price for oil. Public investment in alternative forms of energy, therefore, may be the only available assurance against future rises in OPEC's world price that might go to $15, $20, or even $30 a barrel.

Notes to chapter three

1. For a good account of these conflicts and rivalries, see Malcolm H. Kerr, *The Arab Cold War 1958-1967,* 2d ed. (London: Oxford University Press, 1967).

2. Galileo's apocryphal aside when forced to deny the earth's motion.

3. *Keesing's Contemporary Archives 1975,* p. 27053. The territorial claims on Bahrain and Kuwait, acute until the 1960s, have long since been withdrawn.

4. Indeed, Algerian oil, which remains France's main source, has usually been priced above that of neighboring Libya. When Libya in July 1975 reduced its price, Algeria declined to follow suit—secure perhaps, in its special French connection.

5. See *New York Times,* January 2, 1974, p. 37; cf. above, p. 54.

6. Tucker and Ignotus assume that production could be restored in three months. According to Aramco's President, Frank Jungers (quoted in *Newsweek,* March 31, 1975, p. 48) the time required would be "at least 2 years." For a detailed and dispassionate study of the military, technical, political, and legal aspects of the invasion scenario, see U.S. Library of Congress, Congressional Research Service, *Oil Fields as Military Objectives: A Feasibility Study,* House of Representatives, Committee on International Relations, Special Subcommittee on Investigations, Committee Print (Washington: GPO, 1975). The report suggests that "Jungers' prediction may be overly pessimistic" (p. 17). The authors conclude: "This country could easily defeat OPEC's armed forces in any given locale, while seizing oil fields and facilities, but preserving installations intact would be a chancy proposition under ideal conditions . . .

"... success would largely depend on two prerequisites:

"—Slight damage to key installations.

"—Soviet abstinence from armed intervention.

"Since neither essential could be assured, military operations to rescue the United States (much less its key allies) from an air-tight OPEC embargo would combine high costs with high risks . . . Prospects would be poor, with plights of far-reaching political, economic, social, psychological, and perhaps military consequence the penalty for failure." (Pp. 75, 76.)

7. Tucker, somewhat implausibly, seems to assume that non-Middle Eastern OPEC countries and even Iran might continue production and even exports to the United States ("Oil: The Issue of American Intervention," *Commentary,* p. 27).

8. President Sadat in an interview with a *Newsweek* editor

(published April 9, 1973) boasted openly of the impending use of the "oil weapon" in any future conflict with Israel. For the immediate background, see Robert B. Stobaugh, "The Oil Companies in the Crisis," in Vernon, ed., *The Oil Crisis (Daedalus),* pp. 179-202, esp. 182 ff.

9. Adelman, "Is the Oil Shortage Real?", p. 89.

10. John H. Lichtblau, "Arab Oil and a Settlement of the Middle East Conflict," a paper given at the 24th Annual Conference of the Middle East Institute, Washington, D.C., and reprinted in *PIW,* October 21, 1974, p. 9.

11. Abdulrahman al-Atiqi, quoted in the *Wall Street Journal,* November 29, 1974.

12. Campbell and Caruso, *The West and the Middle East,* p. 47. One of the most perceptive assessments of the psychological attitude of Middle Easterners toward the West remains an article by Albert Hourani, "The Decline of the West in the Middle East," *International Affairs,* January 1953. See also Bernard Lewis, *The Middle East and the West* (Bloomington: Indiana University Press, 1964). Both writings, of course, antedate the recent petroleum revolution.

13. See text, p. 48f, and notes 8-10 of the preceding chapter. OECD forecasts total imports of 21.7 mb/d if the price remains steady at $9 (in 1972 dollars), and 28.9 mb/d if it drops by one-third (*Energy Prospects,* vol. I, p. 12); to these must be added OPEC exports to non-OECD countries, which in 1974 amounted to about 3 mb/d; to assume that these will not increase, as we have done, yields a conservative estimate of total OPEC exports in the 1980s. The other predictions of OPEC exports, in mb/d, are 24.7 (Irving), 29.6 (Morgan), 31 (FNCB), and 31.5 (Levy).

14. OECD forecasts (*Energy Prospects,* vol. I, p. 12) are in millions

of tons of oil per year, which have been converted to mb/d at the standard factor of .020, and augmented by 3 mb/d for exports to non-OECD areas (see preceding note); this yields a figure of 21.5 mb/d for the higher and 33.4 mb/d for the lower price assumption. But the higher of these figures presupposes petroleum exports from the United States in the amount of 1.3 mb/d, an assumption that the authors of the OECD report themselves consider unlikely on second thought (see their footnotes, vol. I, pp. 12 and 11). The First National City Bank forecasts OPEC 1985 exports at 32 mb/d.

15. It would be surprising if a country as fiercely independent as Norway, strongly sympathetic with the Third World, and with 400 years of experience of its own of living under foreign (Danish and Swedish) rule, would not follow OPEC's price lead. Note that the recent discovery of oil and fears that this might benefit the rest of the European Communities more than Norway itself were a major factor in Norway's referendum of 1972 that rejected membership in the EC. Since then, Norway has decided on a slow rate of oil development geared not to the needs of would-be importers but to the development potential of Norway itself; hence the estimates of 4 mb/d from the North Sea, half from the British and half from the Norwegian sector, may be high for the Norwegian share.

When the pipeline that brings the first North Sea oil to the Scottish shore was inaugurated, Prime Minister Harold Wilson reiterated an earlier remark of his, saying, "It was not entirely misplaced humour when I told our friends abroad that a British Minister of Energy will be the chairman of OPEC in the 1980s." (*New York Times,* November 4, 1975.) This presumably meant, at the very least, that Great Britain, too, would follow OPEC's price lead should it ever have any surplus of oil to export.

16. For production figures and estimates of production capacity, see appendix, Tables 5 and 1. Saudi oil minister Sheikh Yamani in April 1975 told a group of American Senators that the industrial countries could cut consumption by at most 3 mb/d, that Saudi Arabia alone

could reduce production by that much and OPEC as a whole by 6 mb/d (*New York Times,* April 19, 1975). Detailed scenarios of production cuts by OPEC countries to 22 mb/d and 19.5 mb/d are given by Mabro, "Can OPEC Hold the Line?" and by Elizabeth Monroe and Robert Mabro, *Oil Producers and Consumers: Conflict or Cooperation* (New York: American Universities Field Staff, 1974), p. 74. For minimum monthly production figures, 1973-75, see Table 7 in the appendix.

17. On the peripheral price movements, cf. note 62 to chapter 1. Headlines such as "OPEC Is Starting to Feel the Pressure," *Fortune,* May 1975, and "Oil-Price Shaving Is Spreading in OPEC; Broad 'Cheating' by Members Seen Next," *Wall Street Journal,* July 18, 1975, are typical of the wishful thinking of the press.

18. *Energy Prospects,* vol. 1, p. 12, places OECD oil imports for the "base case" (i.e., if prices had remained at, or reverted to, 1972 levels in real dollars) at 2014.3 million tons per year (or 40.5 mb/d) for 1980 and 2634.0 million tons per year (or 52.9 mb/d) for 1985.

19. Cf. *New York Times,* March 9, 1975; *Petroleum Press Service,* May 1973, p. 198.

20. See his interview with Oriana Fallaci, "A Sheik Who Hates to Gamble," *New York Times Magazine,* September 14, 1975, p. 19. The preceding sentence seems to misquote Yamani as saying that "we [Saudi Arabia] are limiting ourselves to 3.5 million" barrels a day production. What he must have said is that Saudi Arabia is limiting itself to producing at 3.5 mb/d less than its capacity. Cf. note 16, above.

21. Adelman, *WPM,* pp. 211-15, and "Is the Oil Shortage Real?", p. 88; cf. Penrose, *The Growth of Firms,* p. 239.

22. See, e.g., his letter to the *New York Times,* published October

3, 1974, and the article by Edward Cowan, *ibid.*, September 15, 1975, p. 28.

23. The weaknesses of the sealed bid system are enumerated by Krueger, *The U. S. and International Oil,* p. 187; by John H. Lichtblau, "Uncle Sam Would Be a Weak Oil Bargainer," *New York Times,* April 30, 1975; and "An Oil Importing Bureaucracy," *Wall Street Journal,* editorial, April 23, 1975.

24. Adelman, "Is the Oil Shortage Real?", p. 85; *New York Times,* September 15, 1975; cf. Enders, "OPEC and the Industrial Countries," p. 630.

25. For European taxes on petroleum products in the 1950s and 1960s, see ENI, cited in note 11 to chapter 1; *Pétrole 71;* and Table 8 in the appendix.

26. The latter figure is the refineries' acquisition cost for domestic crude oil for June 1975, as reported by Federal Energy Administration, *Monthly Energy Review,* September 1975.

27. Carroll L. Wilson, "A Plan for Energy Independence."

28. The figures given include both crude oil and natural gas liquids. See American Petroleum Institute, *Reserves of Crude Oil, Natural Gas Liquids in the United States and Canada* for the respective years.

29. *New York Times,* July 25, 1975.

30. Enders, "OPEC and the Industrial Countries," p. 632.

31. *Ibid.*

[FOUR]

Conclusion

OPEC was formed in 1960 to protest a loss of revenue of a few cents per barrel and took fully five years to redress its grievance. Before another five years were up, its members began to escalate their demands alternately for cash and for control. By 1975 the takeover of the production concessions had been completed, and the members' petroleum revenues had risen fiftyfold. A petty contest among auditors in dusty principalities had turned into history's most daring bid for reallocating the world's product. The penny-ante games of the 1960s had become the high finance of the 1970s as OPEC was playing for stakes of hundreds of billions.

Access to raw materials has long been a major theme of international politics. OPEC achieved its successes as oil was replacing coal as the chief fuel of the industrial world. Its growing economic strength had repercussions directly on the relative economic positions of Europe, Japan, and the United States, but also indirectly on the protracted Arab-Israeli conflict and on diplomatic relations among the industrial countries, the communist powers, and the new nations of the Third World.

OPEC proceeded gradually, pragmatically, and yet with increasing boldness, exploring new possibilities in cartel economics and later in international politics. One such new possibility was to found a cartel upon the setting of prices rather than of production quotas. Another, closely related one was the transformation of the zero-sum game between concessionaire companies and host governments into a quiet partnership of benefit to each. Under this arrangement, the tax-paid costs, as periodically adjusted by the governments, set a world-wide floor price; and on the basis of these costs the companies adjust the month-to-month production quotas among members of the cartel. And since the petroleum industry is more vertically integrated than most—from the production (or at least the "offtake") of crude to the retailing of gasoline, fuel oil, and other products—any price increases are smoothly passed on to the ultimate consumer.

The companies themselves are rewarded by secure supplies and by adding their profit margins to a vastly more expensive product. Meanwhile, the petroleum industry's services in developing alternative energy sources, such as offshore oil and perhaps new uses of coal and expanded nuclear power, tend to consolidate its influence in the politics and economics of the home countries. No simple panacea, such as federally administered sealed bids for oil imports into the United States, is likely to dissolve this double nexus.

Proposals for American invasion of one or another oil country seem to us rash and naïve. Economic counterwarfare by the industrial on the oil-producing countries—morality aside—would not produce the intended effect. And drastic measures of curtailing oil consumption at home are likely to hurt the consumer countries long before they would threaten the cohesion of the cartel. All this, in our estimate, adds up to the prospect, over the next five or ten years, not of a sharp break in the price of oil but rather a continuation of present prices (whether or not eroded by world inflation), or even a slight gradual rise. The price of alternatives to oil may plausibly be expected to set a ceiling to OPEC gains in the longer run. Yet there is the danger that, as in the contest between oil and coal in the 1950s, the price of the dearer commodity will have the more magnetic effect than that of the cheaper one. (The recent shift in the concern of consumer-country diplomats from bringing down OPEC's price to setting a floor price for oil imports illlustrates the danger.)

Nevertheless, if OPEC may be earning even larger amounts from its customers than other observers have predicted, it seems to us that OPEC leaders may become adept at spending larger sums sooner than the same analysts have feared. Let us hasten to add that a vast margin of error is inherent in any of the foregoing estimates; and the effect of such errors, whether in the numerator or denominator of a given proportion (such as of OPEC imports to OPEC revenues), will rise exponentially over the years.

In sum, OPEC seems to us, in the light of the preceding analysis, stronger and more durable than many competent observers have allowed. Yet we suggest that the world, nonetheless, can live with OPEC without having to lapse into doomsday fantasies. For the short run, the arrangements worked out by bankers and governments in channeling and rechanneling the vast flood of petrodollars have worked better than was feared in 1974. For the longer run, the problem is one of cumulative transfers of real income from the industrial to the oil-producing countries, transfers that, according to one recent estimate, will attain a rate of 1.7% per annum.[1]

Here Dr. Chenery's analogy of the Marshall Plan is pertinent. The Marshall Plan, of course, was a program of American aid to Europe. But ever since President Truman's Point Four speech of 1951, the same principle of economic assistance has been vastly extended.

For a quarter-century, the rich nations of the world were urged by the Truman and Kennedy administrations, by the United Nations, by liberal economists and political scientists—in sum, by respectable and informed opinion around the globe—to spend at least 1% of their gross national products yearly to help the less fortunate countries of the Third World to help themselves. The liberal target was never met more than part way. But recently fortune's wheel has turned. Some few among the once-poor countries have indeed been helped to help themselves—helped by the industrial world's unquenched thirst for petroleum and by the accidents of geology and international affairs to help themselves to perhaps 1.7% per annum of their customers' product. This time the target has been dramatically oversubscribed (although the rich nations in the last generation have more than doubled their wealth and presumably can afford more); and the selection of recipients of the transfer is accidental and narrow.

Nonetheless, it would be unseemly for former advocates of voluntary foreign aid by the industrial countries to complain now of the wheel's turn, or to take undue alarm at the consequences of OPEC's program of involuntary foreign aid. The only legitimate complainants may be some residents of the "Fourth World" who, having once received less than promised from *anciens riches* nations now must take their chances on the highly selective generosity of the newly rich—or another unexpected turn of the same wheel. The more reason for all three parties to seek constructive occasions for discussing the possibilities of a better world-wide economic order.

OPEC's meteoric rise since the 1960s also confirms what Truman's original formula acknowledged—that the most important transfers are those not of financial resources but of technical skills. OPEC's growing boldness was solidly supported by its leaders' technical competence and shrewd insights into the realities of international political economy. Documents such as the Declaratory Statement of 1968 and Sheikh Yamani's speech of that same year were the culmination of a process of growing sophistication that had begun in Venezuela in the 1940s and spread to Iran and to the Arab countries by the 1960s. This growing competence and sophistication is a just source of pride of the OPEC nations—the same sort of pride that Egyptians felt after 1957 in demonstrating to skeptical Westerners that they could manage large-scale technical and engineering projects such as the traffic through the Suez Canal or the building of the Aswan Dam. For many Westerners who are still caught up in a "bygone view of the capacity of leaders in less-developed countries," [2] and for the American and European press which cannot rid itself of the cliché of caricaturing Middle Easterners as camel-borne nomads, here is a perceptual difficulty—for which, as for all major self-deceptions, there is a heavy price.

For the OPEC nations of the Middle East and elsewhere, thanks to the very success of their leaders as managers of the world's most lucrative cartel, there is now more learning to be done. They must extend their knowledge and sharpen their skills in such fields as large-scale economic development, international investment, urban planning, currency management, and international financial aid. Vast amounts of money and effort are being and must continue to be allocated to enlarging this pool of trained manpower from the political and managerial elites downward. In the Middle East, Kuwait probably is best off, since for years it has attracted a large proportion of the highly literate Arab refugee population from Palestine. But Iran is seeking college teachers from as far away as the United States. The IMF recently seconded one of its high officials, a Pakistani national, to manage the Saudi central bank. Abu Dhabi and other Gulf principalities are importing thousands of construction workers from Pakistan and elsewhere. Such shortages of skill and of manpower are likely to remain characteristic of Middle Eastern oil countries in their quest for rapid industrialization. In real human history there is no such thing as balanced economic growth.[3]

Outside the Middle East, Venezuela, with a literacy rate far higher than any Middle East oil country, has a large headstart on its OPEC associates. For some other countries, such as Nigeria and Indonesia, there may not even be enough of an initial pool of skilled manpower to launch any orderly process of growth. Since these are also the most populous countries, containing more than two-thirds of OPEC's aggregate population, the discrepancy between the needs of the masses and the capabilities of the elite remains large, and the prospects for economic development correspondingly more precarious.

Although OPEC has attracted the widest attention for the challenge it presents to the world's established postcolonial economic order, the challenge for the OPEC elites themselves is to continue to prove that they understand the world's political economy and the mechanism of cartels as well as or better than their erstwhile Western mentors, that they feel as strong a sense of obligation to future generations as the patriots of any other clime, and that with skill and perseverance they may continue to elevate the status of their children among the nations of this world.

NOTES

1. Enders, "OPEC and the Industrial Countries," p. 626.

2. James E. Akins, "The Oil Crisis: This Time the Wolf Is Here," *Foreign Affairs,* vol. 51, no. 3 (April 1973), pp. 462-90, at 473n.

3. See Albert O. Hirschman, *The Strategy of Economic Development* (New Haven: Yale University Press, 1958).

APPENDIX A

Statistical Tables

General Notes:

1) Throughout the tables two dots (. .) mean "not applicable"; three dots (. . .) mean "no data available"; and "0" (or 0.0, etc.) means less than 0.5 (or 0.05, etc.); mb/d = million barrels per day. Totals may not add up exactly because of rounding.

2) On Tables 1, 2, and 6, a heavy step-ladder line separates OPEC member countries from nonmember countries. On the same tables, OPEC totals or averages for 1950 and 1955, placed in brackets, refer to the five countries which in 1960 were to become the founders of OPEC.

3) On Table 9 (and throughout the text) companies are referred to by their current names. They have been previously, or are alternatively, known as follows:

BP = Anglo-Iranian Oil Co., Anglo-Persian Oil Co.
CFP = Compagnie Française des Pétroles
Exxon = Standard Oil Co. of New Jersey, Jersey Standard, Esso
Mobil = Socony, Socony-Vacuum
Shell = The Royal Dutch/Shell Group
Socal = Chevron, Standard Oil Co. of California, Stancal

Table 1. Oil Exporting and Importing Countries: Selected Data, 1950-1975

Country	Producing Since	In OPEC Since	Production, Exports, Imports (million barrels per day)										Capacity	Reserves	Res./Prod.
			1950	1955	1960	1965	1970	1971	1972	1973	1974	1975	mb/d	Bill. bbls	yrs.
			Production												
Saudi Arabia[a]	1938	1960	0.55	0.98	1.31	2.21	3.80	4.77	6.01	7.60	8.48	7.08	10.79	173	80
Iran	1913	1960	0.66	0.33	1.07	1.91	3.83	4.54	5.02	5.86	6.02	5.35	6.60	66	33
Venezuela	1917	1960	1.50	2.16	2.85	3.47	3.71	3.55	3.22	3.37	2.98	2.35	3.00	15	17
Kuwait[a]	1946	1960	0.34	1.10	1.69	2.36	2.99	3.20	3.28	3.02	2.55	2.05	3.29	82	105
Iraq	1928	1960	0.14	0.70	0.97	1.31	1.55	1.69	1.47	2.02	1.87	2.25	2.60	35	41
Qatar	1949	1961	0.03	0.12	0.17	0.23	0.36	0.43	0.48	0.57	0.52	0.44	0.65	6	32
Indonesia	1893	1962	0.13	0.24	0.41	0.48	0.85	0.89	1.08	1.34	1.40	1.31	1.70	15	33
Libya	1961	1962	..	..	..	1.22	3.32	2.76	2.24	2.17	1.52	1.51	2.50	27	68
Abu Dhabi[b]	1962	1967	..	..	..	0.28	0.69	0.93	1.05	1.32	1.41	1.40	1.94	30	58
Algeria	1958	1969	..	..	0.18	0.56	1.03	0.79	1.06	1.10	1.02	0.95	1.00	8	24
Nigeria	1958	1971	..	..	0.02	0.27	1.08	1.53	1.82	2.05	2.25	1.79	2.60	21	36
Ecuador	1918	1973	0.01	0.01	0.01	0.01	0.01	0.01	0.08	0.21	0.18	0.16	0.26	3	39
Gabon	1957	1973	..	..	0.02	0.03	0.11	0.11	0.13	0.15	0.20	0.20	0.21	2	26
Dubai[b]	1969	1974	..	..	..	..	0.07	0.12	0.12	0.21	0.24	0.25	0.30	2	24
Sharjah[b]	1974	1974	..	..	..	..	..	..	..	..	0.02	0.04	0.05	...	...

Table 1, *continued*

			Production, Exports, Imports (million barrels per day)										Capacity	Reserves	Res./Prod.
			1950	1955	1960	1965	1970	1971	1972	1973	1974	1975	mb/d	Bill. bbls	yrs.
OPEC															
Production	..	..	[3.19]	[5.27]	7.89	13.19	22.13	25.08	26.73	30.78	30.77	27.13	37.47	485	49
Exports	..	..	[3.03]	[5.01]	7.50	12.53	21.05	23.66	25.39	29.44	29.54	25.77	..	..	..
Exports as % of World Imports	..	.	[83.0]	[82.3]	83.0	83.6	82.2	84.1	84.1	86.1	88.6	...	..	..	..
World Imports	..	..	3.65	6.09	9.03	14.99	25.60	28.14	30.21	34.19	33.33	...	..	..	..
W. Europe	..	..	1.22	2.40	3.96	7.60	12.94	13.52	14.07	15.41	14.84	...	..	..	..
Japan	..	..	0.04	0.15	0.68	1.72	4.28	4.72	4.82	5.48	5.43	...	..	..	..
U.S.	..	..	0.85	1.25	1.82	2.47	3.42	3.93	4.74	6.26	6.13	6.00	..	..	..
U.S. Production	1859	..	5.41	6.81	7.04	7.81	9.64	9.47	9.44	9.21	8.80	8.37	...	35	11

a. Includes one-half of Neutral Zone Production

b. In 1974, Abu Dhabi's membership was transferred to the United Arab Emirates which includes Dubai and Sharjah

Table 2. Revenues from Petroleum: Selected OPEC Governments and Major Companies, 1950-74
A. Per Barrel Revenues (in Current U.S. $)

	1950	1955	1960	1965	1970	1971	1972	1973	1974
					OPEC Governments				
Saudi Arabia	.632	.821	.750	.832	.883	1.259	1.437	1.789	8.986
Iran	.479	.818	.801	.811	.862	1.246	1.358	2.012	8.397
Venezuela	.597	.807	.892	.956	1.092	1.411	1.719	2.393	10.621
Kuwait	.103	.767	.765	.789	.828	1.197	1.409	1.763	7.734
Iraq	.400	.862	.786	.817	.957	1.415	1.507	2.127	10.149
Qatar	.091	.836	.864	.822	.915	1.264	1.445	1.923	8.465
Indonesia	...	...	..	...	.698	1.040	1.243	2.112	6.521
Libya	..	..	..	.838	1.090	1.786	1.966	2.896	13.893
Abu Dhabi	..	..	..	.325	.920	1.272	1.434	1.621	6.677
Algeria	..	..	...	...	.907	1.268	1.877	2.356	10.571
Nigeria	..	..	...	...	1.093	1.722	1.870	2.797	8.917
OPEC	[512]	[810]	.816	.841	.952	1.365	1.546	2.122	8.931
					Seven Major Companies				
Net Earnings from Petroleum Operations									
Eastern Hemisphere	...	...	.565	.418	.336	.341	.283	.698	...
Western Hemisphere (excluding U.S.)[a]	...	...	.479	.442	.527	.652	.515	...	...
World	...	...	.533	.426	.377	.401	.324	...	...
Payments to Governments									
Eastern Hemisphere	...	...	.708	.764	.865	1.264	1.341	...	...

a. Western Hemisphere earnings in the 1960 column are for 1961.

Table 2, *continued*

B. Annual Revenues (in millions of current U.S. $)

	1950	1955	1960	1965	1970	1971	1972	1973	1974	$ per capita 1974
					OPEC Governments					
Saudi Arabia	113	288	355	655	1,200	2,149	3,107	5,100	27,500	5,000
Iran	91	91	285	522	1,136	1,944	2,380	4,100	17,400	556
Venezuela	331	596	877	1,135	1,406	1,702	1,948	2,800	10,600	911
Kuwait	12	307	465	671	895	1,340	1,657	1,900	7,000	7,527
Iraq	19	207	266	375	521	840	575	1,500	6,800	631
Qatar	1	34	54	69	122	198	255	400	1,600	17,778
Indonesia	...	...	...	...	185	284	429	900	3,000	24
Libya	..	..	..	371	1,295	1,766	1,598	2,300	7,600	3,393
Abu Dhabi	..	..	..	33	233	431	551	900	4,100	19,524
Algeria	..	..	...	...	325	350	700	900	3,700	227
Nigeria	..	..	...	...	411	915	1,174	2,000	7,000	117
OPEC	[565]	[1,489]	2,248	3,798	7,318	11,919	14,374	22,800	96,300	362
					Seven Major Companies					
Net Earnings from Petroleum Operations										
Eastern Hemisphere	...	...	1,101	1,353	1,917	2,236	1,986	5,197	...	
Western Hemisphere (excluding U.S.)[a]	...	...	548	619	825	1,002	769	...	...	
World	...	...	1,649	1,972	2,742	3,238	2,755	...	...	
Payments to Governments										
Eastern Hemisphere	...	...	1,381	2,471	4,938	8,296	9,466	...	...	

a. Western Hemisphere earnings in the 1960 column are for 1961.

Table 3. Calculation of Persian Gulf Crude Oil Prices, 1973-75 (U.S. $ per barrel)

	1/1/1973	10/1/1973	11/1/1973	1/1/1974	1/1/1975
Saudi Arabian 34° gravity oil					
1. Posted price	2.591	3.011	5.176	11.651	11.251
2. Royalty [12.5 percent of (1)]	.324	.376	.647	1.456	2.250
3. Production Cost	.100	.100	.100	.100	.120
4. Profit for tax purposes (1) minus (2 plus 3)	2.167	2.535	4.429	10.095	8.881
5. Tax [55 percent of (4)]	1.192	1.394	2.436	5.552	7.549
6. Government revenue on equity oil (2) plus (5)	1.516	1.770	3.083	7.008	9.799
7. Oil company cost on equity oil (3) plus (6)	1.616	1.870	3.183	7.108	9.919
8. Oil company cost on participation oil	2.330	...	...	10.835	10.460
9. Weighted average cost of equity (7) and participation (8) oil	1.794	...	...	9.344	10.240
10. Weighted government revenue (8) minus (3)	1.694	...	...	9.244	10.120

Table 4. Petroleum Revenues of OPEC Governments, Representative Crudes, 1971-75 (U.S. $ per barrel)

A. Persian Gulf

	Saudi Arabia (Light)	Iran (Light)	Kuwait	Iraq (Basrah)	Qatar (Dukhan)	Abu Dhabi (Murban)	Neutral Zone (Khafji)
1/1/1971	0.99	0.98	0.96	0.93	1.05	1.01	0.68
2/15[a]	1.26	1.25	1.23	1.24	1.32	1.27	0.94
6/1[a]	1.33	1.31	1.29	1.30	1.38	1.29	1.00
1/20/1972[b]	1.45	1.43	1.41	1.42	1.49	1.46	1.11
1/1/1973[a]	1.52	1.50	1.47	1.49	1.55	1.53	1.18
4/1[b]	1.62	1.60	1.57	1.59	1.65	1.63	1.27
6/1[a]	1.70	1.68	1.65	1.67	1.74	1.72	1.35
7/1	1.74	1.72	1.68	1.71	1.78	1.75	1.38
8/1	1.80	1.78	1.75	1.77	1.85	1.82	1.45
10/1	1.77	1.75	1.72	1.74	1.81	1.82	1.42
10/16	3.05	3.02	2.94	3.00	3.15	3.58	2.60
11/1	3.08	3.21	2.97	3.04	3.48	3.62	2.63
12/1	3.00	3.12	2.89	2.95	3.38	3.52	2.55
1/1–6/30/1974	9.27	9.49	9.26	9.30	9.82	10.06	8.93
7/1–9/30	9.37	9.58	9.35	9.39	9.91	10.17	9.02
10/1–10/31	9.69	9.92	9.68	9.72	10.26	10.52	9.31
11/1–12/31	10.09	10.33	10.08	10.12	10.70	10.98	9.64
1/1/1975	9.98	10.26	10.08	10.12	10.62	10.45	9.64
10/1	11.00	11.19	11.09	. . .	11.28	. . .	. . .
Production Cost, 1974	0.16	0.12	0.07	0.15	0.25	0.17	0.45

a. Increases under Tehran Agreement of 2/14/71.

b. Increases under first and second Geneva Agreements.

Table 4, *continued*

B. Other Locations (and "marker" crude for comparison)

	Saudi Arabia (Light)	Saudi Arabia (Light) ex Sidon)	Venezuela (Tia Juana 26°)	Indo-nesia	Libya	Nigeria (Light)	Ecuador (Esmeraldas)
1/1/1971	0.99		1.14				
2/15	1.26						
3/18			1.51				
3/20					2.02		
4/1				2.21			
6/1	1.33						
7/1					2.00		
10/1				2.26	1.99		
1/1/1972			1.60		1.98		
1/20	1.45	1.81			2.16		
4/1		1.80	1.69	2.96	2.14		
7/1					2.13		
8/17							1.36
10/1			1.68				
1/1/1973	1.52	1.79	1.79		2.22	2.02	1.43
3/13			1.96				
4/1	1.61	1.92	2.04	3.73	2.38	2.16	1.63
5/16							1.83
6/1	1.70	2.05			2.52	2.29	
6/22							2.09

Table 4, *continued*

7/1		2.13	2.19		2.62	2.37	
7/2	1.74						
8/1	1.80	2.22	2.44		2.73	2.46	
9/1			2.64				
10/1	1.77	2.23	2.90	4.75	2.86		
10/16	3.05	3.97			5.48		
10/17							3.19
10/20						4.90	
11/1	3.08	4.01	4.47	6.00	5.56	4.95	
11/10							4.54
12/1	3.00	3.90	4.59				
12/15							6.33
1/1/1974	9.27	7.93	8.64	10.80	9.41	8.83	8.78
2/1			9.05				
4/1				11.70			
7/1	9.37		9.38	12.60	9.55		9.21
10/1	9.69				10.33		9.93
11/1	10.09	11.16					
1/1/1975	10.07	(a)	9.58		10.95	10.77	
7/1						10.12	
10/1	11.00		10.35	12.80	11.15	11.31	10.32
Production Cost, 1974	0.16	0.10	0.42	. . .	0.70	0.33	0.45

a. Suspended sales on February 9, 1975.

Table 5. Petroleum Production of OPEC Countries, 1973-75 (Million barrels per day)

	1973				1974				1975			
	Sept.	Oct.	Nov.	Dec.	Jan.–Mar.	Apr.–June	Jul.–Sept.	Oct.–Dec.	Jan.–Mar.	Apr.–June	Jul.–Sept.	Oct.–Dec.–
Saudi Arabia	8.57	7.73	6.27	6.61	7.81	8.87	8.61	8.63	7.05	6.59	7.86	6.80
Iran	5.83	6.02	6.05	6.11	6.15	6.18	5.92	5.88	5.64	5.23	5.68	4.86
Venezuela	3.39	3.38	3.38	3.36	3.23	3.00	2.96	2.83	2.51	2.45	2.32	2.01
Kuwait	3.53	3.09	2.47	2.55	2.84	2.85	2.20	2.31	2.17	2.10	2.26	1.70
Iraq	2.11	1.80	2.15	2.16	1.88	1.82	1.81	1.96	2.01	2.33	2.45	2.23
Qatar	0.60	0.60	0.47	0.46	0.52	0.52	0.52	0.52	0.48	0.44	0.32	0.52
Indonesia	1.42	1.41	1.45	1.45	1.45	1.47	1.43	1.27	1.26	1.19	1.37	1.40
Libya	2.29	2.38	1.77	1.77	1.95	1.82	1.47	1.00	0.96	1.28	2.00	1.79
Abu Dhabi	1.40	1.36	1.17	1.03	1.33	1.62	1.52	1.20	0.91	1.48	1.65	1.55
Algeria	1.10	1.00	0.90	0.90	0.95	1.03	0.90	0.85	0.90	0.80	0.90	0.97
Nigeria	2.14	2.19	2.24	2.26	2.25	2.30	2.20	2.21	1.83	1.59	1.77	1.95
Ecuador	0.12	0.10	0.21	0.14	0.23	0.24	0.11	0.07	0.11	0.16	0.17	0.18
Gabon	0.19	0.19	0.16	0.19	0.17	0.17	0.19	0.19	0.20	0.21	0.20	0.20
Dubai	0.27	0.21	0.14	0.14	0.23	0.24	0.24	0.26	0.24	0.26	0.24	0.27
Sharjah	. .	. .	. .	. .	. .	. .	0.05	0.05	0.04	0.04	0.04	0.04

Table 5, *continued*

	1973				1974				1975			
	Sept.	Oct.	Nov.	Dec.	Jan.–Mar.	Apr.–June	Jul.–Sept.	Oct.–Dec.	Jan.–Mar.	Apr.–June	Jul.–Sept.	Oct.–Dec.
Total OPEC	32.96	31.46	28.83	29.13	30.99	32.13	30.13	29.23	26.31	26.15	29.23	26.47
Participants in 1973/4 Embargo	17.76	16.37	13.19	13.46	15.63	16.95	15.46	14.77	12.71	12.95	15.23	13.60
Non participants	15.20	15.09	15.64	15.67	15.36	15.18	14.67	14.46	13.60	13.20	14.00	12.87
States												
With Production Maxima	19.78	18.54	15.53	15.78	17.68	18.68	17.28	16.49	14.45	14.34	16.41	14.38
Without Maxima	13.18	12.92	13.30	13.35	13.31	13.45	12.85	12.74	11.86	11.81	12.82	12.09
Index numbers (Sept. 1973 = 100)												
Total OPEC	100.0	95.5	87.5	88.4	94.0	97.5	91.4	88.7	79.8	79.3	88.7	80.3
Participants in Embargo	100.0	92.2	74.3	75.8	88.0	95.4	87.0	83.2	71.6	72.9	85.8	76.6
Non participants	100.0	99.3	102.9	103.1	101.1	99.9	96.5	95.1	89.5	86.8	92.1	84.7
With Maxima	100.0	93.7	78.5	79.8	89.4	94.4	87.4	83.4	73.1	72.5	83.0	72.7
Without Maxima	100.0	98.0	100.9	101.3	101.0	102.0	97.5	96.7	90.0	89.6	97.0	91.7

Table 6. Foreign Exchange Reserves of OPEC and Selected Other Countries, 1950-75

(Billions of U.S. $)

											Reserves/Imports (years)		
	1950	1955	1960	1965	1970	1971	1972	1973	1974	1975	1960	1970	1974
OPEC													
Saudi Arabia	...	...	.19	0.73	0.66	1.44	2.50	3.88	14.29	20.43	0.81	0.93	4.11
Iran	0.25	0.21	.18	.24	.21	.62	.96	1.24	8.38	9.69	...	0.13	1.48
Venezuela	.37	.53	.61	.84	1.02	1.52	1.73	2.42	6.53	8.43	0.51	0.51	1.61
Kuwait[a]	...	...	.07	.12	.20	.29	.36	.50	1.40	1.67	0.29	0.32	0.92
Iraq	.12	.29	.26	.24	.46	.60	.78	1.55	3.27	2.81	0.67	0.90	1.71[c]
Qatar	...	...	...	...	...	...	...	...	...	...	...	...	...
Indonesia	.36	.31	.35	...	.16	.19	.57	.81	1.49	1.40	...	0.16	0.40
Libya	...	...	.08	.25	1.59	2.67	2.93	2.13	3.62	2.36	0.47	2.86	1.15
Abu Dhabi	...	...	...	...	...	...	...	...	...	...	...	...	...
Algeria	...	...	...	...	.34	.51	.49	1.14	1.69	1.11	...	0.27	0.46
Nigeria	...	...	.43	.25	.22	.43	.39	.59	5.63	6.20	0.71	0.21	2.06
Ecuador	.04	.03	.04	.05	.08	.07	.14	.24	.35	.26	0.35	0.30	0.37
Gabon	...	...	.01	.02	.01	.03	.02	.05	.07[b]	...	1.25	0.12	0.33[c]

Table 6, *continued*

	1950	1955	1960	1965	1970	1971	1972	1973	1974	1975	Reserves/Imports (years) 1960	1970	1974
OPEC Total	..	..	1.31	2.43	4.64	8.27	10.71	14.55	46.72	54.36	..	..	..
OPEC Index (1971 = 100)	..	..	15.8	29.4	56.1	100.0	129.5	175.9	564.9	657.3	..	..	..
Industrial Countries	...	...	47.10	53.14	65.81	94.15	105.74	115.04	119.91	123.44	..	..	..
United States	22.82	21.75	19.36	15.45	14.49	13.19	13.15	14.38	16.06	16.53	..	..	..
West Germany	.27	3.08	7.03	7.43	13.61	18.66	23.79	33.15	32.40	32.68	..	..	..
United Kingdom	3.44	2.16	3.72	3.00	2.83	6.58	5.65	6.48	6.94	6.39	..	..	..
Japan	.56	1.34	1.95	2.15	4.84	15.36	18.37	12.25	13.52	14.61	..	..	..
Total World[d]	54.55	58.95	60.30	70.03	92.46	130.35	158.08	181.85	217.93	228.69	..	..	..
OPEC % of World	..	..	2.17	3.47	5.02	6.34	6.78	8.00	21.44	23.77	..	..	..

Note: Figures for end of year. 1975 figures are generally for June but otherwise the latest month available. Qatar and United Arab Emirates are not included in OPEC total.

a. Currency Board Holding only; does not include other government-held foreign reserves.
b. November 1974.
c. 1973 data.
d. Excluding U.S.S.R., Eastern Europe, and People's Republic of China.

Table 7. Lowest and Highest Monthly Production of OPEC Countries, 1973-75 (Million barrels per day)

	1973		1974		1975		1973/1975[a]		
	Lowest	Highest	Lowest	Highest	Lowest	Highest	Lowest	Highest	Difference
Saudi Arabia	6.27	8.66	7.52	9.04	5.87	8.41	5.87	9.04	3.17
Iran	5.73	6.11	5.72	6.22	4.76	6.10	4.76	6.22	1.46
Venezuela	3.26	3.43	2.77	3.29	1.77	2.74	1.77	3.43	1.66
Kuwait	2.47	3.77	2.05	2.88	1.58	2.71	1.58	3.77	2.19
Iraq	1.48	2.16	1.66	2.18	1.95	2.55	1.48	2.55	1.07
Qatar	0.46	0.61	0.51	0.52	0.26	0.61	0.26	0.61	0.35
Indonesia	...	...	1.21	1.48	1.10	1.42	1.10	1.48	0.38
Libya	1.77	2.38	0.96	2.03	0.91	2.10	0.91	2.38	1.47
Abu Dhabi	1.03	1.40	1.17	1.65	0.75	1.83	0.75	1.83	1.08
Algeria	0.90	1.19	0.80	1.09	0.80	1.00	0.80	1.19	0.39
Nigeria	1.93	2.26	1.99	2.34	1.56	2.00	1.56	2.34	0.78
Ecuador	...	...	0.06	0.25	0.03	0.21	0.03	0.25	0.22
Gabon	...	...	0.16	0.21	0.20	0.21	0.16	0.21	0.05
Dubai	0.14	0.29	0.18	0.27	0.20	0.28	0.14	0.29	0.15
Sharjah	..	..	0.03	0.06	0.03	0.05	0.03	0.06	0.03
Total	...	...	26.79	33.51	21.77	32.22	21.20	35.65	14.45

a. 1974/75 for Indonesia, Ecuador, Gabon, and Sharjah.

Table 8. Prices and Taxes on Selected Petroleum Products in France, 1950-75

	1950	1955	1960	1965	1970	1971	1972	1973	1974	1975
Gasoline, Paris										
(Francs per litre)										
Pretax price	0.20	0.20	0.26	0.23	0.27	0.27	0.31	0.32	0.71	0.73
Tax	.23	.45	.74	.72	.80	.80	.80	.80	.90	.96
Retail Price	.43	.64	.99	.94	1.06	1.07	1.11	1.12	1.61	1.69
No. 2 Fuel Oil, Atlantic										
(Francs per ton)										
Pretax price	53.43	73.75	106.38	90.74	81.49	92.79	114.02	97.94	223.20	300.64
Tax	5.87	14.15	12.22	10.76	14.61	16.61	20.28	17.56	39.70	53.06
Retail Price	59.30	87.90	118.60	101.50	96.10	109.40	134.30	115.50	262.90	353.70
Tax Rate (percent)										
Gasoline	111.8	228.7	288.2	317.8	300.0	290.5	261.6	246.8	127.7	130.6
Fuel oil	11.0	19.2	11.5	11.9	17.9	17.9	17.8	17.9	17.8	17.7

Table 9. Company Shares of OPEC Petroleum Production, 1961-74 (Million barrels per day)

A. All OPEC Countries, 1961-74

	1961	1965	1970	1971	1972	1973	1974
Exxon	1.98	2.91	3.91	3.98	4.05	3.51	2.51
BP	1.52	2.00	3.87	4.44	4.51	1.94	1.13
Shell	1.37	1.76	2.91	3.13	2.90	1.97	1.49
Gulf	1.15	1.59	2.26	2.50	2.45	1.72	0.85
Texaco	0.70	0.76	1.98	2.27	2.74	2.41	1.63
Socal	0.69	0.97	1.88	2.19	2.62	2.23	1.50
Mobil	0.51	1.14	1.22	1.36	1.48	1.09	0.61
CFP	0.35	0.47	1.11	1.09	0.93	0.63	0.50
Independents	0.63	1.47	3.74	3.37	3.18	2.40	2.22
Government companies	0.06	0.13	0.54	0.99	2.25	13.10	18.30
Total	8.96	13.19	23.41	25.33	27.09	30.99	30.73

Table 9. *continued*

B. Selected OPEC Countries, 1970 and 1974 mb/d

	1970							1974						
	Saudi Arabia	Iran	Venezuela	Kuwait	Iraq	Libya	Nigeria	Saudi Arabia	Iran	Venezuela	Kuwait	Iraq	Libya	Nigeria
Exxon	1.07	0.25	1.69	..	0.18	0.63	..	0.99	..	1.36	..	..	0.08	..
BP	..	1.40	..	1.17	0.37	0.21	0.40	..	..	..	0.46	0.19	..	0.32
Shell	..	0.49	1.08	..	0.37	0.14	0.40	..	..	0.84	..	0.08	..	0.32
Gulf	..	0.25	0.21	1.57	..	..	0.23	..	..	0.16	0.46	..	..	0.17
Texaco	1.07	0.25	0.15	..	..	0.16	0.00	0.99	..	0.11	..	..	..	0.00
Socal	1.07	0.25	0.06	..	..	0.16	0.00	0.99	..	0.05	..	..	..	0.00
Mobil	0.36	0.25	0.12	..	0.18	0.16	0.05	0.33	..	0.06	..	..	0.03	0.11
CFP	..	0.21	..	..	0.37	..	..	..	..	..	..	0.19	..	..
Independents	0.22	0.34	0.37	0.22	0.07	1.85	0.01	0.23	0.23	0.33	0.23	..	0.48	0.11
Government Companies	0.03	0.17	0.05	0.04	..	..	..	4.96	5.79	0.07	1.40	1.53	0.92	1.24
Total	3.80	3.83	3.71	2.99	1.55	3.32	1.08	8.48	6.02	2.98	2.55	1.98	1.52	2.26

Sources and Notes for Tables

Table 1:

OPEC's *Annual Statistical Bulletin 1973* has the dates members began production and joined OPEC, and production data, 1950-73. Production data for 1974 and 1975 are from *Petroleum Intelligence Weekly,* (various issues) as is the "Usable Capacity" information. Petroleum reserves are taken from the *Oil and Gas Journal,* December 30, 1974.

Export data for 1970-72 are from OPEC's *Annual Statistical Bulletin 1972*. According to the same source, OPEC exports, 1967-72, averaged 95% of production. This proportion was used to calculate OPEC exports for other years.

World, Western European, Japanese (1960-74), and United States imports and U.S. production are given according to the annual publication of British Petroleum, *BP Statistical Review of the World Oil Industry 1960, 1969, 1970,* and *1974,* and *API,* January 25, 1976. Japanese imports (1950) are from the United Nations Statistical Office, *World Energy Supplies 1929-1950,* Statistical Papers, Series J, No. 1 (New York: United Nations, 1960). The U.S. Bureau of Mines, *International Petroleum Annual 1965, 1970, 1971, 1972,* and *1973* (each published in March, two years later), gives import data that occasionally differ from the British Petroleum data, especially for Europe and Japan; the difference for the world as a whole ranging from 10,000 to 700,000 barrels per day. Therefore, the reader should consider the figures as approximate.

The increase in OPEC's production and its steady share of world exports are striking. So is the decreasing proportion of U.S. production to imports.

Reserve figures are *published* figures of estimates of "*proven,*" "*recoverable*" reserves and are at best gross approximations. What reserves will be economically "*recoverable*" depends on the future cost of various technologies and the future price of oil—two major sources of uncertainty. Whether high or low estimates are *published* depends in part on the self-interest of the company or government that does the publishing—a company fearful of nationalization may be tempted to understate the reserves, a government eager to enhance its influence in international petroleum politics may be inclined to overstate them. Above all, reserves are "*proven*" by drilling, and hence the figures are likely to be too low wherever the reserves-to-annual-production ratio is high—it rarely is worth the cost to engage in a drilling program to prove that you have, say, 150 years' rather than 75 years' worth of reserves. Conversely, they are likely to become more reliable as the reserves-to-production ratio declines. Even here the reader, of course, should not infer that after the given number of years a country will run "dry." The U.S. reserves-to-production ratio, for example, has been in the range of 8 to 13 years for most of the past half-century—the sort of circumstance that has prompted Adelman *(WPM, passim),* with a hyperbolical, grocery-store metaphor, to refer to proven reserves as the companies' "shelf inventory."

TABLE 2:

Revenue data for 1950 are from Charles Issawi and Mohammed Yeganeh, *The Economics of Middle Eastern Oil* (New York: Praeger, 1962), p. 129; for 1955-72 from

[146]

Petroleum Information Foundation, *Background Information,* Nos. 8 and 16 (1970 and 1973); and for 1973-74, from OECD estimates reported in *The Petroleum Economist,* March 1975, p. 85. Population figures to obtain per capita incomes are taken from United Nations, *Monthly Bulletin of Statistics,* March 1975, pp. 1-5. For 1950, 1973, and 1974 per-barrel revenues for OPEC as a whole have been obtained by dividing the annual revenue figures in part B of the table by annual export figures corresponding to the daily figures given in Table 1.

Data on the seven major companies appear in First National City Bank, *Energy Memo,* October 1970, January 1973, July 1973, January 1974, and January 1975. The seven companies are British Petroleum, Exxon, Gulf, Mobil, Royal Dutch/Shell Group, Standard Oil of California (Socal), and Texaco.

Note that, although the companies' share of revenue from petroleum declined compared to the governments' share, it still generally went up, at times steeply, both per barrel and in the aggregate.

TABLE 3:

[United States,] *Annual Report of the Council on International Economic Policy* (transmitted to Congress, March 1975), p. 158, and *idem* (February 1974), p. 111.

This table may help readers find their way through the maze of "posted prices," "equity" and "participation" oil and "buy-back prices" that bedevilled Persian Gulf oil economics in the early 1970s and is discussed briefly in the text above, p. 26.

TABLE 4:

The data in this table are from the October 15, 1973; April 15, 1974; October 15, 1974; April 15, 1975; and October 15, 1975 editions of Parra, Ramos, and Parra (in cooperation with the *Middle East Economic Survey)*, *International Crude Oil and Product Prices* (Beirut: Middle East Petroleum and Economic Publications, 1973, 1974, 1975). Calculations for some 1975 figures are derived from *PIW* (various issues).

For Persian Gulf crudes, the figures given through 1973 are government take on "equity oil," which was about 75% of all oil sold f.o.b. the Gulf. Private estimates put the actual Saudi "government take" during the first three quarters of 1973 at about $0.08 higher, and from $0.14 to $0.43 higher in the fourth quarter. For 1974 and 1975, the calculations assume a 60-40 participation/equity production split. Calculations by *PIW,* December 23, 1974, of the actual proportions indicate that this assumption overstates the government take by $0.01 to $0.09 per barrel, except Kuwait (as much as $0.31) and Abu Dhabi (as much as $0.27). For 1974, figures in some cases are averages for the periods indicated; the amounts for the first half of the year were imposed retroactively.

For other countries, figures for Indonesia refer to the official selling price, for Libya and Nigeria to equity crude, and for Venezuela and Ecuador to total government revenue. The figures for Saudi Arabia (ex Sidon) are payments to Saudi Arabia only, and do not include pipeline or terminal payments to Jordan, Syria, or Lebanon.

The Persian Gulf crudes generally moved in unison—and they account for two-thirds or more of total OPEC output. Prices of crudes in other locations generally were set later and often with a wider spread. The higher prices in Nigeria indicate a low-sulfur differential, in Venezuela a transportation differential (closeness to the U.S. market), and in Libya both (low sulfur and closeness to Europe). In addition, the Libyan price series indicates the Qaddafi regime's desire to force the upward price movement—although declining production often forced Libya to stand still while others caught up. The Indonesian figures, which include about $3 per barrel of company profit and production cost, are not strictly comparable to the others.

TABLE 5:

Data for Ecuador (1973-74), Gabon (1973-74), Venezuela (September 1974), and Algeria (September 1974) are in *Oil and Gas Journal* (various issues). All other data are taken or calculated from *Petroleum Intelligence Weekly* (various issues, but see especially February 4, 1974; September 30, 1974; November 24, 1975; and each month's final issue).

Notes:

1) Countries listed in order of accession to OPEC.
2) Figures for Saudi Arabia and Kuwait include for each one-half of the production of the Neutral Zone.
3) Abu Dhabi, Dubai, and Sharjah are members of the United Arab Emirates, to which Abu Dhabi's OPEC membership was transferred in 1974.
4) The following states participated in the October 1973 embargo and production cutbacks: Saudi Arabia, Kuwait, Qatar, Libya, Abu Dhabi, Algeria, Dubai.
5) The following states have imposed production maxima: Saudi Arabia, Venezuela, Kuwait, Qatar, Libya, Abu Dhabi (Cf. above, p. 44, n. 61).

The production cuts accompanying the Arab embargo (which were proclaimed in mid-October 1973 and called off in late December) made for a decline by about 25% in the participating countries, for which a slight increase of production elsewhere did not compensate. As a result of the steep price increases of October 1973 and January 1974, world imports (and hence OPEC output) did not again return to the September 1973 levels. Imports rose slightly during the second quarter of 1974, as storage tanks were being filled, and then declined quarter by quarter until mid-1975 while the world's major industrial nations found themselves in a deepening recession. In the second half of 1975, both the major industrial economies and OPEC's oil exports turned up slightly.

TABLE 6:

All data are taken from International Monetary Fund, *International Financial Statistics,* December 1956, October 1966, December 1966, October 1969, July 1975, and August 1975, and *idem, Supplement to 1964/65 Issues.*

Saudi Arabia now ranks (after West Germany and ahead of the United States and Japan) as the second largest holder of exchange reserves. The reserve-to-imports ratio suggests how long OPEC countries could afford to shut down their production without suffering major economic hardship.

TABLE 7:

All data are taken from *Petroleum Intelligence Weekly* (various issues). See also sources for Table 5.

The figures in the last columns suggest by how much

OPEC countries could curtail their production without facing technical difficulties or financial hardship. Conversely, few OPEC countries have reached their capacity production, which for OPEC as a whole is 37.47 mb/d (see Table 1, third column from right).

TABLE 8:

All data are taken from Comité professionnel du pétrole, *Pétrole 74* (Paris: 51, Boulevard de Courcelles, May 1975), pp. 290-97.

Prices are given for January 1 of each year (except for January 11 of 1974).

Figures for 1950 and 1955, given old francs in the source, have been divided by 100 so as to convert them to the new francs in effect since 1960.

Note that in 1970 about three-fourths, and now still well over one-half, of the gasoline price paid by the French motorist goes to his own government.

TABLE 9:

All data for 1961-73 are taken from OPEC, *Annual Statistical Bulletin 1966* (Vienna, 1967), pp. 42-43; *1973* (Vienna, 1974), pp. 38-40; and *1974* (Vienna, 1975), pp. 38-40. Figures for Independents and Government company for Nigeria for 1974 calculated from *PIW*.

The multinational companies listed separately are the so-called Seven Sisters and CFP, in descending order of production. All other foreign companies (U.S., Western European, Japanese) are listed as "independents." "Government companies" refers to the government-owned company, or national company, of each OPEC country, such as Petromin (Saudi Arabia), National Iranian Oil Company (NIOC), Pertamina (Indonesia), Petroven (Venezuela), etc.

The increase in the government companies' share since 1970, and especially in 1974, indicates the progress of nationalization, partial nationalization, and "participation." Note that the multinational companies, under the prevailing "buy-back" provisions (Saudi Arabia, Kuwait, Qatar, Abu Dhabi) or long-term supply agreements (Iran), still distribute nearly all the "participation oil" produced by the government companies. And, hence, it is the multinationals which, in view of world market demand (and prevailing price, quality, and transport differentials), determine how much oil is produced, day by day and month by month, even in the "Government companies" sector—except in Venezuela, Kuwait, and Libya, where government production limits have at times restricted output (cf. p. 44, n. 61).

Note that a majority of cells for 1970 in part B of the table are filled (cf. p. 31). The major companies spread their risk by operating each in three to five of the principal OPEC countries. Conversely, the OPEC countries gained strength from dealing with four to seven of the majors (except Kuwait, which deals only with two). Libya, in addition, gained bargaining strength by having entrusted most of its production to a variety of independents. The buy-back and supply arrangements just mentioned continue this risk-spreading.

APPENDIX B

Chronology

1948	June 28	Kuwait and the American Independent Oil Company (Aminoil) sign a concession agreement for Kuwait's half-interest in the Kuwaiti-Saudi Neutral Zone, which gives Kuwait a 15% interest in the concessionnaire company.
	November 12	Venezuela adopts tax legislation to insure that the government will receive at least 50% of net income from oil production. This becomes the first instance of a "fifty-fifty" arrangement between a host government and oil companies.
1949	February 20	Saudi Arabia and American Pacific Western Oil Company (J. Paul Getty) sign a concession agreement for the Saudi half-interest in

		the Neutral Zone, which gives Saudi Arabia a 25% interest in the company.
1949	September	Venezuela initiates discussions of a possible oil producers' organization with Saudi Arabia, Iran, Iraq, Egypt, Kuwait, and Syria.
1951	January 2	Saudi Arabia and Aramco announce the first "fifty-fifty" profit sharing agreement in the Middle East (signed in early December 1950). Kuwait follows in 1951, Iraq in 1952.
	May 1	Iranian government under Mossadegh adopts the law nationalizing Anglo-Iranian Oil Company (BP). Mossadegh's government is overthrown in August 1953.
1953	June 29	Iraq and Saudi Arabia sign agreement providing for exchange of information on petroleum and consultation on prices.
1954	August	Iran and the Consortium sign agreement, to take effect October 28, that settles the nationalization dispute and embodies the "fifty-fifty" principle.

1957	December 16	Saudi Arabia and Arabian Oil Company (Japanese Petroleum Trading Co.) announce agreement for Saudi offshore interest in the Neutral Zone which gives the government a 56% share of profits. The company signs a similar agreement for Kuwait's offshore interest, providing for a 57% government share.
1959	February 13	The major multinational petroleum companies reduce posted prices for Middle East petroleum.
	April 1	United States government introduces oil import quotas.
	April 16-22	First Arab Petroleum Conference insists that companies cannot unilaterally reduce prices.
1960	May 13	Venezuela and Saudi Arabia call on other producing countries to formulate a "common petroleum policy."
	August 9	The major companies reduce prices further without consulting the governments.
	September 14	The Organization of Petroleum Exporting Countries (OPEC) is

		formed in Baghdad by Iran, Iraq, Kuwait, Saudi Arabia, and Venezuela.
1962	June 7	Fourth OPEC Conference adopts resolution (IV.33), providing that a 50% income tax should be payable in addition to royalty.
1967	June 3-9	Third Arab-Israeli War; all oil ministers of Arab producing states meet at Baghdad and decree an oil boycott (against Great Britain and France), which proves ineffectual. Suez Canal remains closed until 1975.
1968	June 24-25	OPEC's 16th Conference at Vienna adopts "Declaratory Statement of Petroleum Policy in Member Countries."
1970	May 3	TAP line from Saudi Arabia to the Mediterranean interrupted in Syria; from June to December tanker rates rise to all-time highs.
	September 4-October 9	Libya raises its posted prices and increases tax rate from 50% to 55%, retroactive to September 1. Iran and Kuwait follow suit in November.

1970	December 9-12	OPEC's 21st Conference at Caracas establishes 55% as the minimum tax rate and adopts the principle that differences in posted prices should be based only on quality and transportation differentials (Resolution XXI.120), demands that posted prices be changed so as to reflect changes in foreign exchange rates (XXI.122), and instructs its Secretary General to monitor oil company liftings and report on any "discriminatory production policy" (XXI.125).
1971	January 12	Negotiations begin at Tehran between six Gulf producing states and 22 oil companies.
	February 3-4	OPEC's 22nd Conference at Tehran decrees a "total embargo" by member states (except Indonesia) against any company that does not accept the 55% tax rate (Resolution XXII.125).
	February 14	Tehran agreement signed between six Gulf states and 23 oil companies.
	February 24	Algeria nationalizes 51% of French oil concessions.

1971	April 2	Tripoli agreement concluded between oil companies and Mediterranean producers (Libya and Algeria, as well as Saudi Arabia and Iraq for their Mediterranean pipeline throughput). Similar terms are embodied in an agreement between Nigeria and companies operating there, on May 3.
	July 12-13	OPEC's 24th Conference demands "immediate steps towards the effective implementation of the principle of Participation" (Resolution XXIV.135).
	July 31	Venezuela Hydrocarbons Reversion Law requires companies "to cede . . . their unexploited concession areas" by 1974 and "all their residual assets" by 1983.
	September 22	OPEC's 25th Conference appoints Ministerial Committee on participation and directs members (Resolution XXV.140) to negotiate price increases to offset the *de facto* devaluation of the U.S. dollar.
	December 5	Libya nationalizes BP concession in response to British government "collusion" in Iran's occupation of

		several disputed islands in the Persian ("Arab") Gulf.
1972	January 20	Geneva agreement between Middle Eastern oil-producing countries and companies increasing posted prices by 8.49% to offset the decline in the value of the U.S. dollar.
	March 11-12	OPEC's 27th Conference in Resolution XXVII.145 takes note of oil company efforts to undermine its solidarity and threatens unspecified action against companies that "fail to comply with . . . any action taken by a Member Country in accordance with [OPEC] decisions." The implicit reference is to the dispute between Iraq and the multinational concessionnaires.
	June 1	Iraq nationalizes the Iraq Petroleum Company's concession (Kirkuk area) after an eleven-year dispute.
	June 9	OPEC's 28th Conference adopts Resolution XXVIII.146 to prevent companies whose interests were nationalized in Iraq from increasing production elsewhere.

1972	September 30	Libya acquires a 50% interest in two ENI concessions.
	November 6	Saudi Minister Yamani concludes the General Agreement on Participation between Arab Gulf states (Abu Dhabi, Iraq, Kuwait, Qatar, Saudi Arabia) and the companies, providing for 25% government participation on January 1, 1973, rising to 51% by January 1, 1982.
1973	February 28	Iraq and IPC reach agreement on compensation for nationalization.
	March 16	The Shah of Iran announces that the Consortium members have agreed to turn over their assets immediately in return for a long-term supply agreement (which is signed on May 24).
	March 16-17	OPEC's 32nd Conference discusses raising prices to offset further decline in value of the U.S. dollar. A Ministerial Committee is appointed on March 22.
	April 1	OPEC members increase posted prices by 5.7%, following the terms of the Geneva Agreement of January 1972.

1973	April 9	*Newsweek* publishes interview of one of its editors with President Sadat of Egypt, indicating use of "oil weapon" in any future Arab-Israeli war.
	April 18	United States government ends oil import quotas.
	June 1	Six Gulf states, Libya, and Nigeria increase posted prices by 11.9%.
	June 11	Libya nationalizes Bunker Hunt concession; Nigeria acquires 35% participation in Shell-BP concession. In August, Libya nationalizes 51% of the Occidental Petroleum concession and of the Oasis consortium; in September, of nine other companies.
	September 15-16	OPEC's 35th Conference supports Abu Dhabi's efforts to increase posted prices and Libya's nationalizations, and appoints a Ministerial Committee from the Gulf States to "negotiate collectively" an increase in posted prices (Resolutions XXXV.159-160).
	October 6	Fourth Arab-Israeli War begins.
	October 7	Iraq nationalizes Exxon and Mobil shares in Basrah Petroleum Co.

1973	October 8-10	OPEC Ministerial Committee meets with oil company representatives to discuss price increases; no agreement reached.
	October 16-17	Arab oil ministers meeting at Kuwait announce 70% increase in posted prices, and production cuts.
	October 19-20	Libya (10/19) and other Arab oil producers (10/20) announce halting of oil shipments to the United States; during the following week the embargo is extended to the Netherlands.
	November 5	Arab petroleum producers announce production cuts for end of November of 25% below September levels; further cuts of 5% per month are threatened, and a 5% cut for January is announced December 9.
	December 22	Six Persian Gulf producing states (Abu Dhabi, Iran, Iraq, Kuwait, Qatar, Saudi Arabia) raise their posted prices, that for the marker crude going from $5.12 to $11.65 as of January 1, 1974.
	December 25	The Arab oil-producing states announce that exports for January, rather than being cut 5%, will be

		increased 10%; the United States and Netherlands remain on embargo list.
1974	January 7-9	OPEC's 37th Conference decides that posted prices will remain frozen until April 1.
	January 29	Kuwait announces 60% government participation in the BP-Gulf concession; Qatar follows February 20.
	February 11	Washington Energy Conference opens.
	February 11	Libya nationalizes 3 U.S. oil companies that had not agreed to 51% nationalization in September.
	March 18	Arab oil-producing states (except Libya) announce the end of the embargo against the U.S.
	April	U.S. government sets up task force on "Project Independence."
	May 1	U.N. General Assembly at its Sixth Special Session adopts a "Declaration and Programme of Action on the Establishment of a New International Economic Order."

1974	May 18	Nigeria announces 55% government participation in all concessions.
	June 4	Saudi Arabia announces that it will increase its participation in Aramco to 60%. Abu Dhabi and Kuwait follow in September. The increases are retroactive to January 1.
	June 13	IMF establishes its "oil facility."
	July 10-11	Organization of Arab Petroleum Exporting Countries (OAPEC) lifts embargo against the Netherlands.
	July 18	Iran announces that it will acquire a 25% interest in Krupp. In November Kuwait buys one-seventh of Daimler Benz.
	September 13	OPEC's 41st Conference instructs its Secretary General "to carry out a study of supply and demand in relation to possible production controls."
	November 15	International Energy Agency formed at Paris within OECD framework.
	November 15	Saudi Arabia, Qatar, and United Arab Emirates announce a slight

		reduction in posted price and increases in royalty and income tax rates. Earlier (September 6) Saudi Arabia had increased its buy-back price from 93% to 94.9% of posted price.
1974	December 22	Iraq announces plans to develop its production capacity to 3.5 mb/d by late 1975 and to 6 mb/d by 1981.
1975	January 13	*Business Week* publishes Kissinger interview hinting at military action against oil countries in case of "actual strangulation."
	March 4-6	OPEC meeting of heads of state or government at Algiers.
	April 7-15	Preliminary meeting at Paris between oil-exporting, oil-importing, and non-oil Third World countries (Algeria, Saudi Arabia, Iran, Venezuela; European countries, U.S., Japan; India, Brazil, Zaire).
	April 9	24 OECD members decide on Safety Net.
	June 13	World Bank establishes its "Third Window."

1975	September 24	OPEC's 45th Conference announces a 15% increase in government per barrel revenues as of October 1.
	October 28	Venezuela and foreign oil companies agree on nationalization as of January 1, 1976.
	December 1	After protracted negotiations, Kuwait agrees with Gulf and BP on terms of nationalization.
	December 9	Iraq completes nationalization by taking over the BP, CFP, and Shell shares of the Basrah Petroleum Company.
	December 16	Conference on International Economic Cooperation opens at Paris.

APPENDIX C

Declaratory Statement of Petroleum Policy in Member Countries

[Resolution XVI.90, adopted at OPEC's 16th Conference, June 24-25, 1968]

The Conference,

recalling Paragraph 4 of its Resolution I.2;
recognizing that hydrocarbon resources in Member Countries are one of the principle sources of their revenues and foreign exchange earnings and therefore constitute the main basis for their economic development;

bearing in mind that hydrocarbon resources are limited and exhaustible, and that their proper exploitation determines the conditions of the economic development of Member Countries, both at present and in the future;

bearing in mind also that the inalienable right of all countries to exercise permanent sovereignty over their natural resources in the interest of their national development is a universally recognized principle of public law and has been repeatedly reaffirmed by the General Assembly of the United Nations, most notably in its Resolution 2158 of November 25, 1966;

considering also that in order to ensure the exercise of permanent sovereignty over hydrocarbon resources, it is essential that their exploitation should be aimed at securing the greatest possible benefit for Member Countries;

considering further that this aim can better be achieved if Member Countries are in a position to undertake themselves directly the exploitation of their hydrocarbon resources, so that they may exercise their freedom of choice in the utilization of hydrocarbon resources under the most favorable conditions;

taking into account the fact that foreign capital, whether public or private, forthcoming at the request of the Member Countries, can play an important role, inasmuch as it supplements the efforts undertaken by them in the exploitation of their hydrocarbon resources, provided that there is government supervision of the activity of foreign capital to ensure that it is used in the interest of national development and that returns earned by it do not exceed reasonable levels;

bearing in mind that the principal aim of the Organization, as set out in Article 2 of its Statute, "is the coordination and unification of the petroleum policies of Member Countries and the determination of the best means for safeguarding their interests, individually and collectively";

[168]

recommends that the following principles shall serve as basis for petroleum policy in Member Countries.

[1] Mode of Development

1. Member Governments shall endeavour, as far as feasible, to explore for and develop their hydrocarbon resources directly. The capital, specialists and the promotion of marketing outlets required for such direct development may be complemented when necessary from alternate sources on a commercial basis.

2. However, when a Member Government is not capable of developing its hydrocarbon resources directly, it may enter into contracts of various types, to be defined in its legislation but subject to the present principles, with outside operators for a reasonable remuneration, taking into account the degree of risk involved. Under such an arrangement, the Government shall seek to retain the greatest measure possible of participation in and control over all aspects of operations.

3. In any event, the terms and conditions of such contracts shall be open to revision at predetermined intervals, as justified by changing circumstances. Such changing circumstances should call for the revision of existing concession agreements.

[2] Participation

Where provision for Governmental participation in the ownership of the concession-holding company under any of the present petroleum contracts has not been made, the Government may acquire a reasonable participation, on the grounds of the principle of changing circumstances.

If such provision has actually been made but avoided by the operators concerned, the rate provided for shall serve as a minimum basis for the participation to be acquired.

[3] Relinquishment

A schedule of progressive and more accelerated relinquishment of acreage of present contract areas shall be introduced. In any event, the Government shall participate in choosing the acreage to be relinquished, including those cases where relinquishment is already provided for but left to the discretion of the operator.

[4] Posted Prices or Tax Reference Prices

All contracts shall require that the assessment of the operator's income, and its taxes or any other payments to the State, be based on a posted or tax reference price for the hydrocarbons produced under the contract. Such price shall be determined by the Government and shall move in such a manner as to prevent any deterioration in its relationship to the prices of manufactured goods traded internationally. However, such price shall be consistent, subject to differences in gravity, quality and geographic location, with the levels of posted or tax reference prices generally prevailing for hydrocarbons in other OPEC Countries and accepted by them as a basis for tax payments.

[170]

[5] Limited Guarantee of Fiscal Stability

The Government may, at its discretion, give a guarantee of fiscal stability to operators for a reasonable period of time.

[6] Renegotiation Clause

1. Notwithstanding any guarantee of fiscal stability that may have been granted to the operator, the operator shall not have the right to obtain excessively high net earnings after taxes. The financial provisions of contracts which actually result in such excessively high net earnings shall be open to renegotiation.

2. In deciding whether to initiate such renegotiation, the Government shall take due account of the degree of financial risk undertaken by the operator and the general level of net earnings elsewhere in industry where similar circumstances prevail.

3. In the event the operator declines to negotiate, or that the negotiations do not result in any agreement within a reasonable period of time, the Government shall make its estimate of the amount by which the operator's net earnings after taxes are excessive, and such amount shall then be paid by the operator to the Government.

4. In the present context, "excessively high net earnings" means net profits after taxes which are significantly in excess, during any twelve-month period, of the level of net earnings the reasonable expectation of which would have been sufficient to induce the operator to take the entrepreneurial risks necessary.

[171]

5. In evaluating the "excessively high net earnings" of the new operators, consideration should be given to their overall competitive position vis-à-vis the established operators.

[7] Accounts and Information

The operator shall be required to keep within the country clear and accurate accounts and records of his operations, which shall at all times be available to Government auditors, upon request.

Such accounts shall be kept in accordance with the Government's written instructions, which shall conform to commonly accepted principles of accounting, and which shall be applicable generally to all operators within its territory.

The operator shall promptly make available, in a meaningful form, such information related to its operations as the Government may reasonably require for the discharge of its functions.

[8] Conservation

Operators shall be required to conduct their operations in accordance with the best conservation practices, bearing in mind the long-term interests of the country. To this end, the Government shall draw up written instructions detailing the conservation rules to be followed generally by all contractors within its territory.

[172]

[9] Settlement of Disputes

Except as otherwise provided for in the legal system of a Member Country, all disputes arising between the Government and operators shall fall exclusively within the jurisdiction of the competent national courts or the specialized regional courts, as and when established.

[10] Other Matters

In addition to the foregoing principles, Member Governments shall adopt on all other matters essential to a comprehensive and rational hydrocarbons policy, rules including no less than the best of current practices with respect to the registration and incorporation of operators; assignment and transfer of rights; work obligations; the employment of nationals; training programs; royalty rates; the imposition of taxes generally in force in the country; property of the operator upon expiry of the contract; and other such matters.

[11] Definition

For the purposes of the present Resolution, the term "operator" shall mean any person entering into a contract of any kind with a Member Government or its designated agency including the concessions and contracts currently in effect, providing for the exploration for and/or development of any part of the hydrocarbon resources of the country concerned.

Index